Theory and Practice
of Scanning Optical Microscopy

Theory and Practice of Scanning Optical Microscopy

TONY WILSON

Department of Engineering Science, University of Oxford

with contributions from

COLIN SHEPPARD

Department of Engineering Science, University of Oxford

1984

ACADEMIC PRESS

(Harcourt Brace Jovanovich, Publishers)

London   Orlando   San Diego   San Francisco   New York
Toronto   Montreal   Sydney   Tokyo   São Paulo

7188-1281

PHYSICS

ACADEMIC PRESS INC. (LONDON) LTD
24/28 Oval Road,
London NW1

United States Edition published by
ACADEMIC PRESS INC.
(Harcourt Brace Jovanovich, Inc.)
Orlando, Florida 32887

British Library Cataloguing in Publication Data
Wilson, T.
 Theory and practice of scanning optical microscopy.
 1. Microscope and microscopy
 I. Title   II. Sheppard, C. J. R.
 502'.8'2   QH205.2

 ISBN 0-12-757760-2
 LCCCN 83-73235

Filmset by Eta Services (Typesetters) Ltd, Beccles, Suffolk
and printed in Great Britain by
Thomson Litho Ltd., East Kilbride, Scotland

Preface

The use of scanning is well established in electron microscopy, but has only recently become so in other forms of microscopy. After the initial surge of enthusiasm in flying spot light microscopy in the late fifties, interest dwindled until the invention and commercial availability of the laser. This heralded the birth of the scanning optical microscope and of its use, in various guises, in fields as diverse as biology and semiconductor device technology.

This book assumes very little previous knowledge of either optics or microscopy. Its aim is to provide a self-contained account of the major theoretical aspects of scanning optical microscopy as well as its more important applications.

Special thanks are due to colleagues in Oxford who have played a large part in firing interest in this subject: Dr Amarjyoti Choudhury, Mr Ingemar Cox, Dr Julian Gannaway, Dr Douglas Hamilton, Professor Ted Paige, Dr Ebrahim Soleimani, and Dr Laszlo Solymár. I am also indebted to Mrs J. Takacs for the line drawings, and to Catherine Wilson for her editing of the manuscript and her encouragement.

It remains to be said that this book could not have been conceived were it not for the foresight of the late Rudi Kompfner, who introduced the subject to the University of Oxford and whose influence is inevitably to be found in these pages.

Oxford
May 1983

T. WILSON

Acknowledgements

Grateful acknowledgement is made for the various micrographs reproduced in this book. All but two of the micrographs in Chapter 1 were provided by Dr Douglas Hamilton. Figures 4.13, 4.15, and 4.16 are from reference [4.12], Fig. 4.20 is from reference [4.16], and Figs 4.35 and 4.36 are from reference [4.20]. In Chapter 5, Figs 5.2 and 5.3 are from reference [5.2], while Fig. 5.4 is from reference [5.3]. All of the computer-generated micrographs of Chapter 6 were obtained by Mr Ingemar Cox, and special thanks are due to Dr Petráň of the University of Plzeň, Czechoslovakia, for sending the micrographs, shown in Chapter 7, from his direct view confocal microscope. In Chapter 9, Figs 9.8 and 9.9 are from reference [9.12], while Fig. 9.10 is from reference [9.14]. Figures 10.7 to 10.9 in Chapter 10 are from reference [10.16].

T. W.

Contents

Chapter 1

The Scanning Optical Microscope and its Applications

The conventional microscope is an example of a parallel processing system in which the whole area of the specimen is simultaneously imaged either onto a screen or directly onto the retina of the eye. While this can be quite adequate in many cases, the format of the image is not readily suited to subsequent electronic processing, nor can the optical system be easily adapted to take advantage of the various resolution enhancement schemes which we shall discuss in later chapters.

A sequential imaging system provides a much more versatile approach. It may be achieved by scanning a diffraction-limited spot of light relative to a specimen in a raster-type scan. In this way the image is built up point by point, and may be displayed on a TV screen or stored in a computer for future processing. The first example of this kind of light microscope was reported in 1951 by Roberts and Young [1.1]. Their flying spot miscroscope, which was intended for biological studies, used a scanning spot of light from the face of a flying spot scanner tube. The light from the raster was transmitted through conventional microscope optics in reverse, producing a tiny spot of light which was scanned over the sample.

Any radiation passing through the sample fell on a photocell, where it was converted into an electrical signal. The signal was then appropriately amplified and used to modulate the intensity of a TV display scanned in synchronism. The principle advantages of the flying spot microscope are its continuously variable magnification and the electrical form of the image [1.2]. The latter allows particle sizing and counting [1.3–1.5], optical microdensitometry for mass determination [1.6–1.8], image processing for contrast enhancement, resolution improvement by analogue processing, and digital image storage [1.9]. The use of an ultraviolet light source [1.10–1.13,

1

1.22] made the flying spot microscope particularly useful in biological and medical studies.

The later invention of the laser and its incorporation as the light source in a scanning microscope [1.14, 1.15] increased the number of available wavelengths at which the instrument could operate. As a result, imaging with X-rays became possible, as well as imaging in the infrared, which allows the observation of semiconductors in transmission [1.16–1.20]. More recently, the original flying spot has been further developed by using a large area polycrystalline semiconductor laser rather than the phosphor of a cathode ray tube to produce the scanning spot of light [1.21].

A scanning microscope could be constructed, in principle, by scanning a point detector [1.22] across the image field in a conventional microscope, or by using a TV camera in conjunction with a conventional microscope [1.23–1.26] to observe the image. Neither of these schemes is ideal, however. A TV tube, for example, will have variations in sensitivity which will limit the amount of contrast enhancement obtainable. Moreover, damage can result when the entire specimen is bathed in light.

Such problems may be avoided in the scanning optical microscope [1.27–1.31], as only one point on the object is illuminated at a time. This development from the flying spot microscope is shown in Fig. 1.1. A focused laser spot is scanned relative to the object in a TV-like raster. The transmitted or reflected light is collected by a photodetector and the resulting signal is used to modulate the display in the usual fashion.

In the following chapters, we shall discuss the image formation properties of scanning microscopes in great detail. Therefore, in this introductory account of scanning microscopy we shall discuss only a simple model of image formation to indicate that there are two major forms of scanning

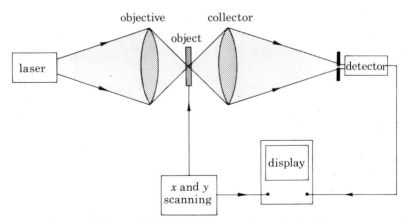

FIG. 1.1. Schematic layout of the scanning optical microscope.

microscope. One of them, the Type 1 arrangement, has imaging properties identical to the conventional instrument, while the other, known as the Type 2 or confocal arrangement, provides greatly improved imaging.

Figure 1.2(a) illustrates the optical system of a conventional microscope, in which the object is illuminated by a patch of light from an extended source through a condenser lens. The object is then imaged by the objective as shown, and the final image is viewed through an eyepiece. In this case the resolution is due primarily to the objective lens, while the aberrations of the

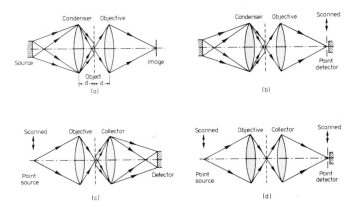

FIG. 1.2. The optical arrangements of various forms of scanning optical microscopes. (a) Conventional microscope. (b) Type 1a scanning microscope. (c) Type 1b scanning microscope. (d) Type 2 or confocal scanning microscope.

condenser are unimportant. A scanning microscope using this arrangement could then be realised by scanning a point detector through the image plane so that it detects light from one small region of the object at a time, thus building up a picture of the object point by point (Fig. 1.2(b)). The arrangement of Fig. 1.2(c), using a second point source and an incoherent detector, has the same imaging properties as the microscope of Fig. 1.2(b) and the conventional microscope if the roles of the two lenses are exchanged. This is the arrangement of the Type 1 scanning microscope. The point source illuminates one very small region of the object, while the large area detector measures the power transmitted by the collector lens. The arrangement shown in Fig. 1.2(d) is a combination of those in Fig. 1.2(b) and (c). Here the point source illuminates one very small region of the object, and the point detector detects light only from the same area. An image is built up by scanning the source and detector in synchronism.

In this configuration we see that both lenses play equal parts in the imaging. We might expect that as two lenses are employed simultaneously to image the object, the resolution will be improved; and this prediction is borne out by both calculation and experiment. This arrangement has been named a Type 2 or confocal scanning microscope. The term confocal is used to indicate that both lenses are focused on the same point on the object. In practical arrangements, however, it is often more convenient to scan the object rather than the source and detector together.

This resolution improvement may at first seem to contravene the basic limits of optical resolution. It may be explained, however, by a principle described by Lukosz [1.32] which states that resolution may be improved at the expense of field of view. The field of view can be increased, however, by scanning. One way of taking advantage of Lukosz's principle is simply to place a very small aperture extremely close to the object [1.33, 1.34]. The resolution is now determined by the size of the hole rather than the radiation. This scheme has been successfully demonstrated at microwave frequencies [1.34], but there are such severe practical difficulties in locating a small enough aperture close enough to the object that the scheme has not been applied at optical frequencies.

This does not mean, however, that we cannot take advantage of the principle at optical frequencies. If, instead of using a physical aperture in the focal plane, we use the back projected image of a point detector in conjunction with the focused point source, we have a confocal scanning microscope. This, then, is a practical arrangement which gives superior resolution at optical frequencies. As an alternative to a point detector we may use a coherent detector, as in the scanning acoustic microscope; or a heterodyning method, as described later in this chapter.

The confocal microscope was first described by Minsky [1.35]. He recognised the arrangement's important depth discrimination properties, which allow optical sectioning of a thick translucent object. Light from the specimen is focused through a small aperture, thus ensuring that information is obtained only from one particular level of the specimen.

The confocal microscope behaves as a coherent optical system [1.36] in which the image of a point object is given by the product of the image by the lens before the object (the objective) with that formed by the lens behind the object (the collector). This results in a sharpened image of a single point with extremely weak outer rings, which gives rise to images without artefacts. Brakenhoff [1.37] has already obtained a resolution of 100 nm using a He–Cd laser (wavelength 325 nm) and immersion lenses.

By using detectors with different sensitivity geometries, it is possible to produce images which depend on specific object properties. Dekkers and de Lang [1.38], for example, used a large area detector split into two halves to obtain a differential phase contrast image by displaying the difference signal from the two halves of the detector. The detector used by Koester [1.39], on the other hand, imaged one strip of the object at a time, and the complete image was built up by scanning. This provided Type 1 imaging in one direction and Type 2 in the other.

We have previously suggested the confocal arrangement as a method of improving the resolution of a scanning microscope. It is also reasonable to ask if the resolution of a conventional microscope could be similarly enhanced. This question will be discussed in more detail in Chapter 7 in connection with Petráň's microscope [1.40, 1.41], where a Nipkow wheel is used for scanning. In this instrument the light passes through the wheel both before and after striking the specimen, producing an image which can be viewed directly through an eyepiece.

We have now discussed the theoretical aspects of the scanning optical microscope, but we have said little about the major practical problem of deciding how to scan. There are essentially two different kinds of scanning, which have been achieved by various methods in practical instruments. The alternatives are either to scan a focused light beam across a stationary object, or to scan the object mechanically across a stationary spot. In the first case, in which the scanning can be very fast, many whole pictures can be built up per second, so that rapid changes within objects may be observed. Mechanical scanning, however, produces undistorted images of very high quality, as illustrated in the reflection micrographs of Fig. 1.3(a) [1.42]. Figure 1.3(b) shows the same specimen imaged in a conventional microscope for comparison. Some contrast enhancement has been used in the scanning micrograph to show up the different phases more clearly. This electronic contrast enhancement facility allows the observation of weak detail and precludes the need to stain biological specimens, thus eliminating the risk of

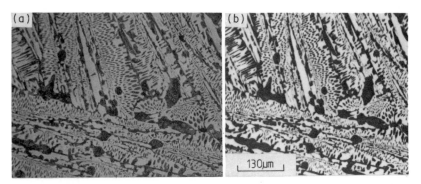

FIG. 1.3. Ledeburite eutectic cast-iron etched with 2% Nital examined in (a) conventional optical microscope; (b) scanning optical microscope using a He–Ne laser (wavelength 6328 nm).

killing or altering living cells in the staining process. A further example is shown in Fig. 1.4, which illustrates the superior resolution of the confocal imaging mode.

A further advantage of mechanical scanning is that the optical path is stationary, which means the lens design is considerably simplified. For example, in the video disc player (which is actually a scanning optical microscope!), a single element plastic moulded lens with two aspheric surfaces has been used [1.43]. Alternatively, we could take advantage of the relaxed off-axis aberration requirements to design a special lens to achieve a

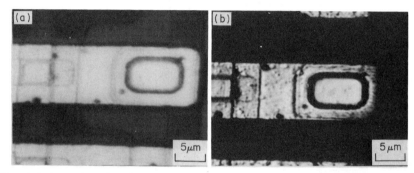

FIG. 1.4. A portion of a microcircuit taken in (a) a conventional microscope and (b) a confocal scanning microscope with the same lenses and laser, illustrating the resolution improvement attained with the confocal arrangement.

longer working distance, higher numerical aperture, or shorter operating wavelength than usual. The advantages of the stationary optical system are further illustrated in Fig. 1.5, where the resolution has been maintained over the entire image.

Mechanical scanning also makes interference scanning microscopy easy to achieve without the necessity of matched optics [1.44], although Hundley [1.45] has managed to construct such an instrument using the flying spot method. In addition, by adding and subtracting the object beam and reference beam, the real and imaginary parts of the object transmittance may be determined [1.46–1.48].

The first alternative, scanning the laser beam, has generally been used when only modest resolution has been required, and usually involves the mechanical movement of a scanning mirror, or acousto-optic beam deflector [1.49], or a weak lens [1.50] in the optical path. Applications where scanning the beam is employed include systems for scanning semiconductor wafers or hybrid microcircuits for dust particles or defects by bright- or dark-field microscopy [1.51–1.55], or by observation of the resultant electrical effects

[1.56], or by fluorescent radiation [1.57]. Commercial wafer scanners based on these principles have also been developed [1.58].

The use of scanning in microscopy allows the use of a wide range of imaging mechanisms which cannot be exploited in conventional microscopy. These modes generally rely on light input to produce some observed effect. An example is the optical beam induced contrast (OBIC) method for imaging electronic properties of semiconductor devices [1.59], in which the laser

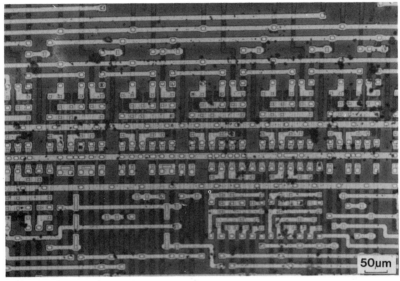

FIG. 1.5. A low magnification micrograph taken with a mechanically scanned optical microscope, illustrating that the resolution is maintained across the entire image.

beam excites carriers to produce a current in an external circuit. Such an image shows up dislocations, grain boundaries, and other defects (Fig. 1.6). Other modes include fluorescence and luminescence microscopy, which give information about the chemical structure and the electronic band structure [1.60]. Figure 1.7(a) shows an OBIC image of a GaAs light emitting diode, while Fig. 1.7(b) shows the image of the light emitted by the diode itself [1.61]. The emitted light image shows very clearly that this particular diode is not uniformly efficient, but emits only around the edge of its active area. This is very useful information which cannot easily be found from the OBIC image, and certainly cannot be seen from a reflected light image.

If fluorescence microscopy is undertaken by scanning an illuminating spot relative to the specimen, the resulting resolution is limited by the input rather than the longer fluorescent wavelength. A pulsed laser may be used to produce periodic heating, and images of thermal properties are formed either by collecting the emitted infrared radiation [1.62], by observing the resulting thermal expansion by interferometry [1.63], or by collecting the emitted sound waves [1.64]. The incident radiation may be used to produce photoemitted electrons, which are then used to produce an image [1.65] giving information about the structure of the surface.

There are many more interesting techniques which arise from the use of a laser beam in scanning microscopy. For example, we can build a resonant microscope [1.66, 1.44] by placing the object inside a resonant cavity similar to a laser cavity. This is practical only if the object is mechanically scanned,

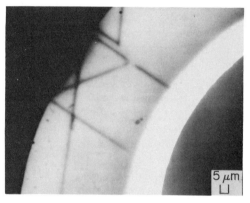

FIG. 1.6. An OBIC image formed by monitoring the emitter base current in a silicon transistor. The dislocation lines are clearly visible.

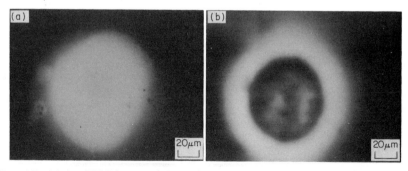

FIG. 1.7. (a) An OBIC image of the active area of a GaAs light-emitting diode. (b) The same diode, but here the image has been formed by monitoring the light emitted by the diode itself.

as the alignment of the cavity mirrors must not be disturbed. Resonant microscopy results in a multiple-beam interference system which, in principle, allows extremely fine variations in height or thickness to be made visible, while using a high numerical aperture for high lateral resolution. The method also results in contrast enhancement, as a weakly absorbing object has a considerable effect on the power circulating in the cavity.

Alternatively, if the laser beam is divided in two and the frequency of one beam is changed slightly, we can construct a heterodyne microscope. One beam travels through the object, the two beams are then recombined, and the detection system is tuned to the beat frequency [1.67]. One result is that phase information is imaged. The detector also detects only the component of the object beam which has the same phase front as the reference beam. Thus, if the latter is a diverging spherical wave, only light from the object which appears to come from the centre of the sphere will be detected. In this way a synthetic lens is produced [1.68, 1.69], which in principle could be used to image with X-rays. Sawatari [1.70] has described a heterodyne microscope in which a real lens and a synthetic lens were used to form a confocal system. If both beams travel through the object slightly displayed in relation to each other, a differential interference microscope is formed [1.71].

The laser has also proved useful in biological microscopy, as it allows selective destruction of specimens. This procedure provides information about specimen structure and function [1.72, 1.73], while the resulting products may be analysed by mass spectrometry [1.74].

Further, a laser allows the investigation of the nonlinear optical properties of an object. For example, images have been formed from the second harmonic produced within the object itself [1.75]. The quantity of harmonic produced depends on the crystal structure and orientation of the object. The harmonic microscope also exhibits a depth descrimination property similar to that in the confocal microscope [1.76]. This arises because the harmonic power is proportional to the square of the input power, so that an appreciable amount of harmonic is formed only in the region of focus. The resolution in harmonic microscopy is increased in relation to that attainable with the primary frequency [1.77], as the squaring sharpens the beam in the lateral direction. It is important to note that because the harmonic radiation need only be collected and not focused, the smallest usable wavelength is determined by the optics available for the primary light. A wide range of other nonlinear optical effects may also be used, including Raman scattering or two-photon fluorescence. In this way, information may be obtained about the energy levels and hence the chemical structure of the object.

References

[1.1] J. Z. Young and F. Roberts (1951). *Nature* **167**, 231.
[1.2] F. Roberts and J. Z. Young (1952). *Proc. IEE* **99**, Pt IIIA, 747.
[1.3] P. N. Slater (1959). *J. Opt. Soc. Am.* **49**, 562.
[1.4] W. E. Tolles and H. P. Mansberg (1962). *Ann. N.Y. Acad. Sci.* **97**, 516.
[1.5] M. J. Eccles, B. D. McCarthy, D. Proffitt and D. Rosen (1976). *J. Microsc.* **106**, 33.
[1.6] H. Noara (1951). *Science* **114**, 279.
[1.7] H. C. Box and H. Freund (1959). *Rev. Sci. Instrum.* **30**, 28.
[1.8] J. J. Freed and J. L. Engle (1962). *Ann. N.Y. Acad.* **97**, 412.
[1.9] P. H. Bartels, R. A. Buchroeder, D. W. Hillman, J. A. Jones, D. Kessler, R. M. Shoemarker, R. V. Shack, D. Turner and D. Vukobratovic (1988). *Analyt. Quant. Cytol. J.* **3**, 55.
[1.10] P. O'B. Montgomery, F. Roberts and W. A. Bonner (1956). *Nature* **117**, 1172.
[1.11] P. O'B. Mongomery (1962). *Ann. N.Y. Acad. Sci.* **97**, 491.
[1.12] G. Z. Williams and R. G. Neuhauser (1962). *Ann. N.Y. Acad. Sci.* **97**, 358.
[1.13] W. A. Bonner (1962). *Ann. N.Y. Acad. Sci.* **97**, 408.
[1.14] P. Davidovits and M. D. Egger (1969). *Nature* **223**, 831.
[1.15] P. Davidovits and M. D. Egger (1971). *Appl. Opt.* **10**, 1615.
[1,16] B. Sherman and J. F. Black (1970). *Appl. Opt.* **9**, 802.
[1.17] G. M. Ovcharenko, N. F. Soboleva and O. K. Shabashev (1975). *Sov. J. Opt. Technol. (USA)* **42**, 588.
[1.18] H. H. Pattee Jr. (1953). *J. Opt. Soc. Am.* **43**, 61.
[1.19] P. Horowitz and J. A. Howell (1951). *Science* **114**, 356.
[1.20] M. Luukkala (1974). *Electron. Lett.* **10**, 481.
[1.21] O. V. Bogdankevich, V. J. Djukov, S. A. Beljnev, S. I. Gavrikov and L. N. Nevorova (1980). *IEEE. J. Quant. Electron.* **OE-16**, 129.
[1.22] E. M. Deeley (1955). *J. Scient. Instrum.* **32**, 263.
[1.23] V. K. Zworykin, L. E. Flory and R. E. Shrader (1952). *Electronics* **24**, 150.
[1.24] V. K. Zworykin and L. E. Flory (1952). *Electron. Engng.* **71**, 40.
[1.25] V. K. Zworykin and F. L. Hatke (1957). *Science* **126**, 805.
[1.26] V. K. Zworykin and C. Berkley (1962). *Ann. N.Y. Acad. Sci.* **97**, 364.
[1.27] R. C. Mellors and R. Silver (1951). *Science* **114**, 356.
[1.28] C. J. R. Sheppard (1977). *IEEE. J. Quant. Elect.* **QE-13**, 49D.
[1.29] T. Wilson (1980). *Appl. Phys.* **22**, 119.
[1.30] C. J. R. Sheppard (1980). *Electron. Power*, February 1980.
[1.31] G. J. Brakenhoff, P. Blom and P. Barends (1979). *J. Microsc.* **117**, 219.
[1.32] W. Lukosz (1966). *J. Opt. Soc. Am.* **56**, 1463.
[1.33] C. W. McCutchen (1967). *J. Opt. Soc. Am.* **57**, 1190.
[1.34] E. A. Ash and G. Nicholls (1972). *Nature* **237**, 510.
[1.35] M. Minsky, U.S. Patent 3013467, Microscopy Apparatus, Dec. 19, 1961 (Filed Nov. 7, 1957).
[1.36] C. J. R. Sheppard and T. Wilson (1980). *Optik* **55**, 331.
[1.37] G. J. Brakenhoff, J. S. Binnerts and C. L. Woldringh (1980). *In* "Scanned Image Microscopy" (Ed. E. A. Ash). Academic Press, London and New York.
[1.38] N. H. Dekkers and H. de Lang (1974). *Optik* **41**, 452.
[1.39] C. J. Koester (1980). *Appl. Opt.* **19**, 1749.
[1.40] M. D. Egger and M. Petráň (1967). *Science* **157**, 305.

[1.41] M. Petráň, M. Hadravsky, M. D. Egger and R. Galambos (1968). *J. Opt. Soc. Am.* **58**, 661.
[1.42] T. Wilson, D. K. Hamilton, P. J. Shadbolt and B. Dodd (1980). *Met. Sci.* April, 1980, 144.
[1.43] G. Bouwhuis and P. Burgstede (1973). *Phillips Tech. Rev.* **33**, 186.
[1.44] G. J. Brakenhoff (1979). *J. Microsc.* **117**, 233.
[1.45] L. L. Hundley (1962). *Ann. N.Y. Acad. Sci.* **97**, 514.
[1.46] C. J. R. Sheppard and T. Wilson (1980). *Phil. Trans. R. Soc. (Lond.)* **295**, 513.
[1.47] T. Wilson (1979). *J. Opt. Soc. Am.* **18**, 3764.
[1.48] D. K. Hamilton and C. J. R. Sheppard (1982). *Optica Acta* **29**, 1573.
[1.49] J. Lekavich, G. Hrbek and W. Watson (1970). EOSD Conference, New York. (See Conference Proceedings, p. 650.)
[1.50] T. O. Casperson and G. M. Lomakka (1962). *Ann. N.Y. Acad. Sci.* **97**, 449.
[1.51] W. J. Patrick and E. J. Patzner (1973). *J. Electrochem. Soc.* **120**, 97c.
[1.52] D. R. Oswald and D. F. Munro (1974). *J. Elect. Mat.* **3**, 225.
[1.53] A. D. Gara (1981). *Electron. Test*, May 1981, 60–70.
[1.54] R. E. Hines (1982). *Electron. Engng.*, August 1982.
[1.55] W. J. Alford, R. D. Vanderneut and V. J. Zaleckas (1982). *Proc. IEEE* **70**, 641.
[1.56] T. H. diStefano and J. J. Cuomo (1977). *Appl. Phys. Lett.* **30**, 351.
[1.57] W. D. Johnston and B. I. Miller (1973). *Appl. Phys. Lett.* **23**, 192.
[1.58] R. Iscoff (1982) *Semicond. Int.* 39, November 1982.
[1.59] C. J. R. Sheppard, J. N. Gannaway, D. Walsh and T. Wilson (1980). *In* "Microcircuit Engineering" (Eds H. Ahmed and W. C. Nixon). Cambridge University Press, Cambridge.
[1.60] J. F. Black, C. J. Sumers and B. Sherman (1972). *Appl. Opt.* **11**, 1553.
[1.61] T. Wilson, J. N. Gannaway and P. Johnson (1980). *J. Microsc.* **118**, Pt III, 309.
[1.62] G. Busse (1980). *In* "Scanned Image Microscopy" (Ed. E. A. Ash). Academic Press, London and New York.
[1.63] S. Ameri, E. A. Ash and C. R. Petts (1981). *Electron. Lett.* **17**, 337.
[1.64] H. K. Wickramasinghe, R. C. Bray, V. Jipson, C. F. Quate and J. L. Salcedo (1978). *Appl. Phys. Lett.* **33**, 923.
[1.65] W. J. Baxter (1977). *JTEVA* **5**, 243.
[1.66] C. J. R. Sheppard and R. Kompfner (1978). *Appl. Opt.* **17**, 2879.
[1.67] A. Korpel and R. L. Whitman (1969). *Appl. Opt.* **8**, 1577.
[1.68] Y. Fujii (1972). National Convention of Institute of Electronics and Electrical Co-munications Engineers of Japan.
[1.69] Y. Fujii, H. Takimoto and T. Igarishi (1981). *Opt. Commun.* **38**, 90.
[1.70] T. Sawatari (1973). *Appl. Opt.* **12**, 2768.
[1.71] L. J. Laub (1972). *J. Opt. Soc. Am.* **62**, 737.
[1.72] N. A. Peppers (1965). *Appl. Opt.* **4**, 555.
[1.73] M. Bessis (1970). *Biol. Med. Phys.* **13**, 209.
[1.74] D. Glick (1981). *In* "Recent Advances in Quantitative Hystochemistry and Cytochemistry". Hans Huber, Bern.
[1.75] J. N. Gannaway and T. Wilson (1979). *Proc. R. Microsc. Soc.* **14**, 170.
[1.76] J. N. Gannaway and C. J. R. Sheppard (1978). *Opt. Quant. Electron.* **10**, 435.
[1.77] T. Wilson and C. J. R. Sheppard (1979). *Opt. Acta* **26**, 761.

Chapter 2

Introduction to the Theory of Fourier Imaging

In order to appreciate the fundamental limitations of the resolution and image formation properties of optical microscopes, it is necessary to begin by discussing the foundations of diffraction theory. We shall then carry on to discuss the most important component of an optical imaging system—the lens—and finally combine our findings in a consideration of the role of Fourier analysis in the theory of coherent and incoherent image formation.

2.1 Kirchhoff diffraction and the Fresnel approximation

We take as our starting point the Kirchhoff diffraction formula, which is the mathematical interpretation of Huygens' principle. For paraxial optics the electromagnetic field may be expressed as a scalar field amplitude. If we are concerned only with radiation of angular frequency ω this may be written as

$$\bar{U} = \text{Re}\,\{U\,e^{j\omega t}\}, \tag{2.1}$$

where U is the complex amplitude and $\text{Re}\,\{\ \}$ denotes the real part. Kirchhoff's diffraction formula [2.1] gives the amplitude in the plane x_2, y_2 in terms of the distribution in the plane x_1, y_1 (Fig. 2.1) as

$$U_2(x_2, y_2) = \int\int\limits_{-\infty}^{+\infty} \frac{1}{j\lambda R}\,U_1(x_1, y_1)\exp\,(-jkR)\,dx_1\,dy_1, \tag{2.2}$$

where k is the wavenumber, given by $k = 2\pi/\lambda$, and λ is the wavelength. This expression assumes that U_1 is slowly varying compared to the wavelength, and that both U_1 and U_2 are only appreciable in a region around the optic axis which is small compared to the axial distance z. According to Huygens'

12

principle each element of the wavefront U_1 may be considered to give rise to a spherical wave with strength Proportional to U_1. The double integral then represents a summation over all elements of the wavefront.

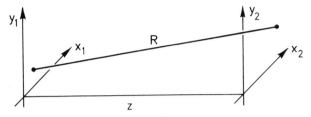

FIG. 2.1. Diffraction geometry.

If we impose a more rigid condition on the maximum values of x_1, y_1 and x_2, y_2, we may replace the distance R in the denominator by z and expand the R in the exponent by the binomial theorem to give the Fresnel approximation

$$U_2(x_2, y_2) = \frac{\exp(-jkz)}{j\lambda z} \int\int_{-\infty}^{+\infty} U_1(x_1, y_1) \exp$$

$$-\frac{jk}{2z} \{(x_2 - x_1)^2 + (y_2 - y_1)^2\} \, dx_1 \, dy_1. \tag{2.3}$$

It should be noted that the assumption that U_1 is slowly varying is necessary for paraxial optics, as a quickly varying amplitude will result in diffraction through large angles.

2.2 The Fraunhofer approximation

If z is large compared to the maximum of x_1 and y_1, the Fresnel approximation to the diffraction integral may be used. However, if the more stringent condition that

$$z \gg \tfrac{1}{2}k(x_1^2 + y_1^2)_{\max} \tag{2.4}$$

is also satisfied, we may make the further approximation of neglecting the terms involving x_1^2 and y_1^2, to give the Fraunhofer approximation

$$U_2(x_2, y_2) = \frac{\exp(-jkz)}{j\lambda z} \exp -\frac{jk}{2z}(x_2^2 + y_2^2)$$

$$\times \int\int_{-\infty}^{+\infty} U_1(x_1, y_1) \exp \frac{jk}{z}(x_1 x_2 + y_1 y_2) \, dx_1 \, dy. \tag{2.5}$$

The condition (2.4) may be written in terms of the Fresnel number N defined as

$$N = \pi(x_1^2 + y_1^2)_{max}/\lambda z, \qquad (2.6)$$

so that if $N \ll 1$ the Fraunhofer form of the diffraction integral may be used.

We now introduce the two-dimensional Fourier transform $\tilde{U}(m, n)$ of $U(x, y)$ which we define as

$$\tilde{U}(m, n) = \int\int_{-\infty}^{+\infty} U(x, y) \exp 2\pi j(mx + ny) \, dx \, dy; \qquad (2.7)$$

the inverse relationship is

$$U(x, y) = \int\int_{-\infty}^{+\infty} \tilde{U}(m, n) \exp -2\pi j(mx + ny) \, dm \, dn. \qquad (2.8)$$

Then we can write (2.5) in the form

$$U_2(x_2, y_2) \exp \frac{jk}{2z}(x_2^2 + y_2^2) = \frac{\exp(-jkz)}{j\lambda z} \tilde{U}_1(x_2/\lambda z, y_2/\lambda z), \qquad (2.9)$$

The exponential term on the left-hand side of equation (2.9) shows that a Fourier transform relationship between the original and diffracted field is satisfied on a spherical surface (to the paraxial approximation) centred on the axial point of the x_1, y_1 plane.

We shall now give three examples of Fraunhofer diffraction which we shall find important later.

2.2.1   *The rectangular aperture*

We define the rectangular function rect (x) by [2.1]

$$\begin{aligned} \text{rect } x &= 1, & |x| &< \tfrac{1}{2} \\ &= 0. & |x| &> \tfrac{1}{2} \end{aligned} \qquad (2.10)$$

The Fourier transform of rect (x/a) is a sinc (am), the sinc function being defined as

$$\text{sinc } \xi = \frac{\sin(\pi\xi)}{\pi\xi}. \qquad (2.11)$$

If an aperture in an opaque screen is illuminated with a plane wave, then, providing the dimensions of the aperture are large compared to the wavelength (as they must be for our paraxial approximation), the amplitude after the aperture is simply the product of the illuminating amplitude and the transmission of the aperture. This amounts to saying that the presence of the aperture does not affect the distribution in amplitude before the aperture, which corresponds to the Kirchhoff boundary conditions. Thus if a rectangular aperture is illuminated with a uniform plane wave, the amplitude U_1 is given by a rect function. The diffracted amplitude U_2 for a rectangular aperture of dimensions a and b may thus be written

$$U_2(x_2, y_2) \exp \frac{jk}{2z} (x_2^2 + y_2^2) = \frac{\exp(-jkz)}{j\lambda z} ab \operatorname{sinc}\left(\frac{ax_2}{\lambda z}\right) \operatorname{sinc}\left(\frac{by_2}{\lambda z}\right).$$

(2.12)

Defining the angles θ, ϕ as $\sin^{-1}(x_2/z)$, $\sin^{-1}(y_2/z)$, we may write for the diffracted intensity, which is merely the modules squared of the diffracted amplitude,

$$I_2(\theta, \phi) = \left(\frac{ab}{\lambda z}\right)^2 \operatorname{sinc}^2\left(\frac{a \sin \theta}{\lambda}\right) \operatorname{sinc}^2\left(\frac{b \sin \phi}{\lambda}\right).$$

(2.13)

2.2.2 The circular aperture

If a function of two coordinates x, y is radially symmetric such that it is a function of r only then its two-dimensional Fourier transform is also radially symmetric and may be written as a Fourier–Bessel (or Hankel) transform

$$\tilde{U}(\rho) = \int_0^\infty U(r) J_0(2\pi\rho r) 2\pi r \, dr,$$

(2.14)

where J_n is a Bessel function of the first kind of order n. If the original amplitude is that of an evenly illuminated circular aperture radius a, written as circ (r_1/a), where the circ function is defined

$$\begin{aligned} \operatorname{circ}(r) &= 1, & r < 1 \\ &= 0, & r > 1, \end{aligned}$$

(2.15)

the diffracted amplitude is given by

$$U_2(r_2) \exp\left(\frac{jkr_2^2}{2z}\right) = \frac{\exp(-jkz)}{j\lambda z} \pi a^2 \left[\frac{2J_1(2\pi r_2 a/\lambda z)}{2\pi r_2 a/\lambda z}\right].$$

(2.16)

Here we have made use of the integral

$$\int_0^1 J_0(2\pi\rho r) 2\pi r \, dr = \frac{J_1(2\pi\rho)}{\rho} = \pi\left[\frac{2J_1(2\pi\rho)}{2\pi\rho}\right].$$

(2.17)

The diffracted intensity may thus be written

$$I_2(\theta) = \left(\frac{\pi a^2}{\lambda z}\right)^2 \left[\frac{2J_1(2\pi a \sin \theta/\lambda)}{2\pi a \sin \theta/\lambda}\right]^2,\tag{2.18}$$

where $\sin \theta$ is r_2/z.

2.2.3   The annular aperture

For an evenly illuminated annular aperture, outer and inner radii a and γa respectively, the amplitude is given by (using 2.16)

$$U(r_2)\exp\left(\frac{jkr_2^2}{2z}\right) = \frac{\exp(-jkz)}{j\lambda z}\,\pi a^2$$

$$\times\left\{\left[\frac{2J_1(2\pi r_2 a/\lambda z)}{2\pi r_2 a/\lambda z}\right] - \gamma^2\left[\frac{2J_1(2\pi r_2 \gamma a/\lambda z)}{2\pi r_2 \gamma a/\lambda z}\right]\right\}.\tag{2.19}$$

In the limiting case of a thin annulus given by $(1 - \gamma) = \varepsilon$, with ε small, the amplitude is

$$U(r_2)\exp\left(\frac{jkr_2^2}{2z}\right) = \frac{\exp(-jkz)}{j\lambda z}\int_0^\infty \delta(r_1 - a)J_0(2\pi r_1 r_2/\lambda z)2\pi\varepsilon r_1\,dr_1,\tag{2.20}$$

where δ is the Dirac delta function, giving

$$U(r_2)\exp\left(\frac{jkr_2^2}{2z}\right) = \frac{\exp(-jkz)}{j\lambda z}\,2\pi\varepsilon a J_0(2\pi r_2 a/\lambda z).\tag{2.21}$$

2.3   The thin lens

We shall now consider the effects of a thin lens, that is a lens which is so thin that the rays do not experience a significant displacement upon traversing it. The lens produces a phase change proportional to the optical path, given by the line integral of the refractive index along the ray. Restricting our analysis to the quadratic approximation, we now consider a perfect thin lens, by which we mean a lens producing the quadratic phase change required to collimate a spherical wave diverging from a point at distance f from the lens into a plane wave. By considering equation (2.9), we see that the phase change produced by the lens must be such as to multiply the amplitude by the factor

$$t(x, y) = \exp\frac{jk}{2f}(x^2 + y^2).\tag{2.22}$$

It may be shown that a thin lens constructed from a dielectric medium with two spherical surfaces gives the required phase variations [2.1].

A real lens also has a finite physical size. The effect of this may be taken into account by introducing the pupil function of the lens $P(x, y)$, which is unity within the pupil and zero outside. In general, the pupil function can be a varying complex function of position to account for absorption in the lens, reflection at the surfaces, or variations which are purposely introduced.

Let us now illuminate the lens with a plane wave of unit strength. The amplitude just after the lens may be written

$$U_1(x_1, y_1) = P(x_1, y_1) \exp \frac{jk}{2f} (x_1^2 + y_1^2). \tag{2.23}$$

The exponential factor represents a spherical wave convergent on the point at distance f beyond the lens. The amplitude in the plane at distance z away may be calculated using equation (2.3), to give

$$U_2(x_2, y_2) = \frac{\exp(-jkz)}{j\lambda z} \int\int\limits_{-\infty}^{+\infty} P(x_1, y_1) \exp \frac{jk}{2f} (x_1^2 + y_1^2)$$

$$\times \exp \frac{-jk}{2z} \{(x_2 - x_1)^2 + (y_2 - y_1)^2\} \, dx_1 \, dy_1. \tag{2.24}$$

Multiplying out the squared brakets, the terms in x_1^2, y_1^2 will be seen to cancel if z is equal to f. Then

$$U_2(x_2, y_2) = \frac{\exp(-jkf)}{j\lambda f} \exp \frac{-jk}{2f} (x_2^2 + y_2^2)$$

$$\times \int\int\limits_{-\infty}^{+\infty} P(x_1, y_1) \exp \frac{jk}{f} (x_1 x_2 + y_1 y_2) \, dx_1 \, dy_1. \tag{2.25}$$

The integral is the Fourier transform of the pupil function, which we denote by $h(x_2, y_2)$. Thus we can write for the intensity in the focal plane of the lens

$$I_2(x_2, y_2) = |h(x_2, y_2)|^2 / \lambda^2 f^2. \tag{2.26}$$

We are often concerned with radially symmetric pupils. Then the two-dimensional Fourier transform may be written as a Fourier–Bessel transform and in the focal plane

$$U_2(r_2) = \frac{\exp(-jkf)}{j\lambda f} \exp\left(\frac{-jkr_2^2}{2f}\right) \int\limits_0^\infty P(r_1) J_0\left(\frac{2\pi r_1 r_2}{\lambda f}\right) 2\pi r_1 \, dr_1. \tag{2.27}$$

If the lens has a circular pupil radius a, then the amplitude is identical to that written in equation (2.16) for Fraunhofer diffraction by a circular aperture. It is convenient to introduce a normalised optical coordinate v, defined by

$$v = kr_2 \sin \alpha \approx 2\pi r_2 a/\lambda f, \tag{2.28}$$

where $\sin \alpha$ is the numerical aperture of the lens (assuming imaging in air). Then we can write for the focal amplitude

$$U_2(v) = -jN \exp(-jkf) \exp\left(\frac{-jv^2}{4N}\right)\left[\frac{2J_1(v)}{v}\right], \tag{2.29}$$

where N is the Fresnel number given by

$$N = \pi a^2/\lambda f. \tag{2.30}$$

If N is large then the condition

$$N \gg v^2/4 \tag{2.31}$$

ensures that the quadratic phase variation in equation (2.29) is negligible for reasonable values of v. The intensity is thus proportional to $[2J_1(v)/v]^2$, which is plotted in Fig. 2.2.

For a pupil function of the form of a thin annulus of fractional thickness ε,

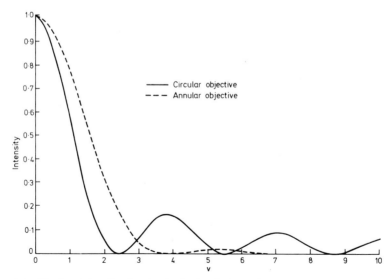

FIG. 2.2. The intensity distribution in the focal plane of a circular lens and an annular lens.

equation (2.21) gives for the amplitude in the focal plane

$$U_2(v) = -2jN\varepsilon \exp(-jkf) \exp\left(\frac{-jv^2}{4N}\right) J_0(v). \tag{2.32}$$

The intensity is now proportional to $J_0^2(v)$, which is also plotted in Fig. 2.2. The central spot is now narrower, but the strength of the outer rings is seen to be increased.

2.4   The effect of defocus

We have considered the amplitude in the focal plane of a lens illuminated by a plane wave, and now discuss the amplitude in a plane a distance δz from the focal plane. The squared terms in x_1, y_1 in equation (2.24) no longer cancel, and we can thus write for the radially symmetric case

$$U_2(r_2) = \frac{\exp(-jkz)}{j\lambda z} \exp\left(\frac{j\pi r_2^2}{\lambda z}\right)$$

$$\times \int_0^\infty P(r_1) \exp\left\{\frac{jkr_1^2}{z}\left(\frac{1}{f} - \frac{1}{z}\right)\right\} J_0\left(\frac{2\pi r_1 r_2}{\lambda z}\right) 2\pi r_1 \, dr_1. \tag{2.32}$$

We can regard the integral as the Fourier–Bessel transform of the product of the pupil function and the complex exponential which is present as a result of the defocus. This product may be thought of as a generalised pupil function, which is complex to account for the aberration of the wavefront at the pupil.

For a circular pupil we can write, introducing $\rho = r_1/a$,

$$U_2(v) = -jN \exp(-jkf) \exp\left(\frac{-jv^2}{4N}\right)$$

$$\times \int_0^1 2 \exp\left\{\frac{jk\rho^2 a^2}{z}\left(\frac{1}{f} - \frac{1}{z}\right)\right\} J_0(v\rho)\rho \, d\rho. \tag{2.34}$$

The wavefront aberration rises to its greatest value at the edge of the pupil where it is equal to $\frac{1}{2}a^2(1/f - 1/z)$ wavelengths. If we define the normalised optical coordinate u by

$$u = ka^2\left(\frac{1}{f} - \frac{1}{z}\right), \tag{2.35}$$

the amplitude is

$$U_2(u, v) = -jN \exp(-jkz) \exp\left(\frac{-jv^2}{4N}\right)$$

$$\times \int_0^1 2 \exp(\tfrac{1}{2}ju\rho^2)J_0(v\rho)\rho \, d\rho. \tag{2.36}$$

If $z = f + \delta z$ with δz small

$$u \approx k \, \delta z a^2/f^2 \approx 4k \, \delta z \sin^2(\alpha/2) \tag{2.37}$$

and u is linearly related to the distance from the focal plane. Then the maximum wavefront aberration is $2 \, \delta z \sin^2(\alpha/2)$.

Along the optic axis we obtain for the amplitude

$$U_2(u, 0) = -jN \exp(-jkz) \exp\left(\frac{ju}{4}\right)\left[\frac{\sin(u/4)}{u/4}\right]. \tag{2.38}$$

Or for the intensity

$$I(u, 0) = N^2\left[\frac{\sin(u/4)}{u/4}\right]^2. \tag{2.39}$$

In general the intensity may be written

$$I(u, v) = N^2[C^2(u, v) + S^2(u, v)], \tag{2.40}$$

where $C(u, v)$ and $S(u, v)$ are defined as [2.2]

$$\left.\begin{aligned}
C(u, v) &= \int_0^1 2 \cos(\tfrac{1}{2}ju\rho^2)J_0(v\rho)\rho \, d\rho, \\[2em]
S(u, v) &= \int_0^1 2 \sin(\tfrac{1}{2}ju\rho^2)J_0(v\rho)\rho \, d\rho.
\end{aligned}\right\} \tag{2.41}$$

These integrals may be evaluated numerically or expressed in terms of Lommel functions. The behaviour of the intensity in the focal region is illustrated in Fig. 2.3, which shows contours of constant intensity, normalised to unity at the focal point. The lines $u = v$ correspond to the shadow edge given by geometrical optics for the paraxial case.

If the pupil function is a thin annulus, the integral in equation (2.32) may be evaluated directly, to give for the amplitude

$$U_2(u, v) = -2jN\varepsilon \exp(-jkz) \exp\left(\frac{-jv^2}{4N}\right) \exp(\tfrac{1}{2}ju)J_0(v). \tag{2.42}$$

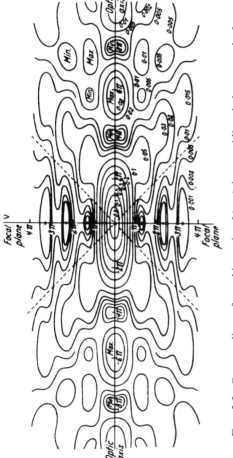

FIG. 2.3. Contour lines of equal intensity, $I(u, v)$, in a meridional plane near the focus of a converging spherical wave diffracted at a circular aperture. The dotted lines represent the boundary of the geometrical shadow. (After reference [2.2].)

The important feature here is that the intensity variation with distance from the optic axis is independent of the value of u within the range of the Fresnel approximation, that is the depth of focus is exceedingly large. As the beam propagates the radiation diffracts away from the axis, but power is simultaneously diffracted inwards from the strong outer rings. A beam with intensity distribution given by $J_0^2(v)$ will propagate without spreading as a result of this dynamic equilibrium: it is a mode of free space.

The imaging properties of lenses and mirrors with annular aperture have been the subject of considerable interest since the work of Airy [2.3] in 1841. In the annular lens the central peak is sharpened but at the expense of increasing the strength of the outer bright rings (Fig. 2.2). The intensity in the focal plane and along the optic axis for an annulus of finite width has been calculated by Steward [2.4, 2.5] who also showed [2.5] that the intensity distribution along the optic axis is stretched out relative to that of a circular lens, that is, the depth of field is increased. This increased depth of field, however, is unfortunately not useful for examining extended objects in the conventional microscope [2.6], as the increase in brightness in the outer diffraction rings results in a loss of contrast, and an n-fold increase in focal depth involves an n-fold loss of light. Because a laser is used as a light source in scanning microscopy this latter point, however, is not a serious drawback in this case.

2.5   Coherent imaging

Let us now assume we have a transparency which is sufficiently thin that it may be described completely by a complex amplitude transmittance $t(x, y)$, of which the variations in modulus represent the variations in absorption in traversing the transparency, whereas the variations in phase account for the optical path travelled. If this is illuminated with an axial plane wave of unit strength the amplitude after the transparency is similarly $t(x, y)$. If the transparency is placed at a distance d_1 in front of a lens (Fig. 2.4), the

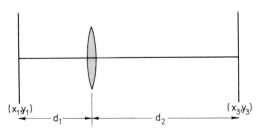

FIG. 2.4. The image formation geometry.

amplitude $U_2(x_2, y_2)$ immediately behind the lens is found by applying the Fresnel diffraction formula and multiplying by the pupil function and phase factor for the lens, to give

$$U_2(x_2, y_2) = P(x_2, y_2) \exp \frac{jk}{2f}(x_2^2 + y_2^2) \frac{\exp(-jkd_1)}{j\lambda d_1}$$

$$\times \int\int_{-\infty}^{+\infty} t(x_1, y_1) \exp \frac{-jk}{2d_1} \{(x_2 - x_1)^2 + (y_2 - y_1)^2\} \, dx_1 \, dy_1.$$

$$(2.43)$$

The amplitude $U_3(x_3, y_3)$ in a plane at a distance d_2 behind the lens is then given by a further application of equation (2.3)

$$U_3(x_3, y_3) = -\frac{\exp\{jk(d_1 + d_2)\}}{\lambda^2 d_1 d_2} \int\int\int\int_{-\infty}^{+\infty} P(x_2, y_2) t(x_1, y_1)$$

$$\times \exp \frac{-jk}{2d_1}\{(x_2 - x_1)^2 + (y_2 - y_1)^2\}$$

$$\times \exp \frac{-jk}{2d_2} \{(x_3 - x_2)^2 + (y_3 - y_2)^2\}$$

$$\times \exp \frac{jk}{2f}(x_2^2 + y_2^2) \, dx_1 \, dy_1 \, dx_2 \, dy_2 \qquad (2.44)$$

$$= -\frac{\exp\{-jk(d_1 + d_2)\}}{\lambda^2 d_1 d_2} \int\int\int\int_{-\infty}^{+\infty} P(x_2, y_2) t(x_1, y_1)$$

$$\times \exp \frac{-jk}{2d_1}(x_1^2 + y_1^2)$$

$$\times \exp \frac{-jk}{2d_2}(x_3^2 + y_3^2) \exp\left\{\frac{-jk}{2}\left(\frac{1}{d_1} + \frac{1}{d_2} - \frac{1}{f}\right)(x_2^2 + y_2^2)\right\}$$

$$\times \exp jk\left[x_2\left(\frac{x_1}{d_1} + \frac{x_3}{d_2}\right) + y_2\left(\frac{y_1}{d_1} + \frac{y_3}{d_3}\right)\right] dx_1 \, dy_1 \, dx_2 \, dy_2.$$

$$(2.45)$$

If the condition known as the lens law

$$\frac{1}{d_1} + \frac{1}{d_2} = \frac{1}{f} \qquad (2.46)$$

is satisfied, and furthermore with

$$d_2 = M d_1, \tag{2.47}$$

we obtain

$$U_3(x_3, y_3) = - \frac{\exp\{-jkd_1(1 + M)\}}{\lambda^2 M d_1}$$

$$\times \int\limits_{-\infty}^{+\infty}\!\!\int\!\!\int\!\!\int P(x_2, y_2)\, t(x_1, y_1) \exp \frac{-jk}{2d_1}(x_1^2 + y_1^2)$$

$$\times \exp \frac{-jk}{2Md_1}(x_3^2 + y_3^2)$$

$$\times \exp \frac{jk}{d_1}\left[x_2\!\left(x_1 + \frac{x_3}{M}\right) + y_2\!\left(y_1 + \frac{y_3}{M}\right)\right] dx_1\, dy_1\, dx_2\, dy_2.$$

$$\tag{2.48}$$

Performing the integral in x_2, y_2 we have

$$U_3(x_3, y_3) = - \frac{\exp\{-jkd_1(1 + M)\}}{\lambda^2 M d_1^2} \exp \frac{-jk}{2Md_1^2}(x_3^2 + y_3^2)$$

$$\times \int\limits_{-\infty}^{+\infty}\!\!\int t(x_1, y_1) \exp \frac{-jk}{2d_1}(x_1^2 + y_1^2) h\!\left(x_1 + \frac{x_3}{M}, y_1 + \frac{y_3}{M}\right) dx_1\, dy_1,$$

$$\tag{2.49}$$

where

$$h(x, y) = \int\limits_{-\infty}^{+\infty}\!\!\int P(x_2, y_2) \exp \frac{jk}{d_1}(x_2 x + y_2 y)\, dx_2\, dy_2 \tag{2.50}$$

is the Fourier transform of the pupil function, as introduced in equation (2.26). Suppose now that our object consists of a single bright point in an opaque background, so that

$$t(x_1, y_1) = \delta(x_1)\, \delta(y_1). \tag{2.51}$$

Then the amplitude is a constant times $h(x_3/M, y_3/M)$, and the latter is called the amplitude point spread function or impulse response of the optical system. The distance x_3/M represents a distance in the object plane, and M is the linear magnification of the image. The intensity is given by the modulus squared of $h(x_3/M, y_3/M)$, again multiplied by a constant, and this is known as the intensity point spread function. For a circular pupil the intensity is (following equations 2.16, 2.29)

$$I(v) = \left[\frac{2J_1(v)}{v}\right]^2,\qquad(2.52)$$

where the normalised coordinate v is given by

$$v = 2\pi r_3 a / \lambda d_1 \qquad(2.53)$$

and we have normalised the intensity to unity on the optic axis. Equation (2.52) represents the Airy disc, as shown in Fig. 2.2

If the lens law (2.46) is not satisfied then for small departures from the focal plane the amplitude point spread function is as given by the previous section on the effects of defocus. For a circular pupil we have

$$h(u, v) = C(u, v) + jS(u, v),\qquad(2.54)$$

where we have normalised to unity for $u = v = 0$, and u is given by

$$u = ka^2 \left(\frac{1}{f} - \frac{1}{d_1} - \frac{1}{d_2}\right).\qquad(2.55)$$

Introducing $x' = x_1 + x_3/M$, $y' = y_1 + y_3/M$ in equation (2.49) we have

$$U_3(x_3, y_3) = -\frac{\exp\{-jkd_1(1 + M)\}}{\lambda^2 M d_1^2} \exp\frac{-jk}{2Md_1}(x_3^2 + y_3^2)$$

$$\times \int\limits_{-\infty}^{+\infty}\!\!\int t\left(x' - \frac{x_3}{M}, y' - \frac{y_3}{M}\right)$$

$$\times \exp\frac{-jk}{2d_1}\left\{\left(x' - \frac{x_3}{M}\right)^2 + \left(y' - \frac{y_3}{M}\right)^2\right\}h(x', y')\,\mathrm{d}x'\,\mathrm{d}y'.$$

$$(2.56)$$

For an imaging system of reasonable quality the spread function falls off quickly, so that x', y' are small. The exponential terms in x'^2, y'^2 and $x'x_3$, $y'y_3$ can therefore be replaced by unity to give

$$U_3(x_3, y_3) = -\frac{\exp\{-jkd_1(1 + M)\}}{\lambda^2 M d_1^2} \exp\frac{-jk}{2Md}\left(1 + \frac{1}{M}\right)(x_3^2 + y_3^2)$$

$$\times \int\limits_{-\infty}^{+\infty}\!\!\int t(x_1, y_1)h(x_1 + x_3/M, y_1 + y_3/M)\,\mathrm{d}x_1\,\mathrm{d}y_1.$$

$$(2.57)$$

The integral is the convolution of the object transmittance with the point spread function, the M's resulting in a magnification M in the image, and the positive sign in the argument of the spread function corresponding to an inverted image. Of the two complex exponential terms in (2.57) the first is a constant phase term which therefore does not affect the image.

The second phase factor in (2.57) represents a spherical phase variation which may also be neglected if we are concerned with small changes in x_3, y_3. In most of the following, however, we are interested in optical systems where the optical beam travels along the axis, in which case there is no phase variation to worry about.

The intensity is clearly given by

$$I_3(x_3, y_3) = \frac{1}{\lambda^4 M^2 d_1^4} \left| \int\int_{-\infty}^{+\infty} t(x_1, y_1) h(x_1 + x_3/M, y_1 + y_3/M) \, dx_1 \, dy_1 \right|^2.$$

$$(2.58)$$

2.6   Imaging of line structures in coherent systems

In the previous section we derived a general expression for the image in a coherent imaging system, and also considered the image of the single point object. An important class of objects are those in which the transmittance is a function of one direction only, let us say $t(x_1)$. Using equation (2.57) the image is, disregarding the phase-terms,

$$U_3(x_3, y_3) = \frac{1}{\lambda^2 M d_1^2} \int\int_{-\infty}^{+\infty} t(x_1) h_1\left(x_1 + \frac{x_3}{M}, y_1 + \frac{y_3}{M}\right) dx_1 \, dy_1. \quad (2.59)$$

Considering the integral y_1, we see that

$$\int_{-\infty}^{+\infty} h\left(x_1 + \frac{x_3}{M}, y_1 + \frac{y_3}{M}\right) dy_1 = \int_{-\infty}^{+\infty} h\left(x_1 + \frac{x_3}{M}, y_1\right) dy_1, \quad (2.60)$$

with the result that, as we might expect, the image is independent of the coordinate y_3. The integral may be written in terms of the pupil function, using equation (2.50) to give

$$\int_{-\infty}^{+\infty} h(x_1, y_1)\, dy_1 = \int\int\int_{-\infty}^{+\infty} P(x_2, y_2) \exp\frac{jk}{d_1}(x_1 x_2 + y_1 y_2)\, dx_2\, dy_2\, dy_1$$

$$(2.61)$$

$$= \int_{-\infty}^{+\infty} P(x_2, 0) \exp\left(\frac{jk x_1 x_2}{d_1}\right) dx_2 = g(x_1). \qquad (2.62)$$

The image may thus be written

$$U_3(x_3, y_3) = \frac{1}{\lambda^2 M d_1^2} \int_{-\infty}^{+\infty} t(x_1) g\left(x_1 + \frac{x_3}{M}\right) dx_1, \qquad (2.63)$$

which is the convolution of th object transmittance and $g(x_1)$. The quantity $g(x_1)$ is called the line spread function, and is the amplitude image of a bright line.

If the pupil is radially symmetric, the line spread function is given by equation (2.62) as the one-dimensional Fourier transform of $P(r_2)$, as compared with the point spread function, which is the Fourier–Bessel, or two-dimensional Fourier, transform of $P(r_2)$. For a lens with $|x| \leqslant a$

$$g(x_1) = \int_{-a}^{+a} \exp\left(\frac{2\pi j x_1 x_2}{\lambda d_1}\right) dx_2$$

$$= 2a\left(\frac{\sin v}{v}\right) \qquad (2.64)$$

where (equation 2.28)

$$v = k x_1 a / d_1. \qquad (2.65)$$

This is plotted in Fig. 2.5, where it is normalised to unity at the origin. This result may also be derived by direct integration of the point spread function as in (2.61). For a thin annular pupil, on the other hand, we have

$$g(x_1) = \int_{-\infty}^{+\infty} \{\delta(x_2 - a) + \delta(x_2 + a)\} \exp\left(\frac{2\pi j x_1 x_2}{\lambda d_1}\right) dx_2 \qquad (2.66)$$

$$= 2 \cos v, \qquad (2.67)$$

which again is shown normalised in Fig. 2.5. The side-lobes are now as strong as the main lobe, and hence imaging of such an extended object with a thin annular pupil is completely useless. The problem is that in the spread function of the annular pupil, the intensity in successive outer rings does not decay to zero.

Another object of great importance is the step object, which has a transmittance defined by

$$t(x_1) = 0, \qquad x_1 < 0, \atop = 1, \qquad x_1 > 0. \Big\}$$

(2.68)

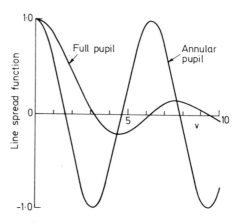

FIG. 2.5. The coherent line spread function for both a circular and an annular lens.

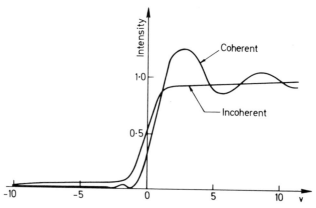

FIG. 2.6. The intensity image of a straight-edge object for both coherent and incoherent systems.

The image may be calculated for a circular pupil using (2.63) to give

$$U_3(x_3) \propto \int_v^\infty \left(\frac{\sin v}{v}\right) dv, \tag{2.69}$$

$$\propto \frac{1}{2} - \frac{1}{\pi} \operatorname{Si}(v), \tag{2.70}$$

where Si is a sine integral and (2.70) is normalised to unity for large negative v. The intensity is plotted in Fig. 2.6, showing that the image exhibits fringes. For a thin annular pupil, the image integrates to a constant.

2.7 The coherent transfer function

An alternative approach to imaging is to consider the object in terms of its spatial frequencies. A periodic object may be resolved into a Fourier series, whereas for a non-periodic object we must use a Fourier transform. The Fourier transform of the object transmittance is

$$T(m, n) = \int\int_{-\infty}^{+\infty} t(x, y) \exp 2\pi j (mx + ny) \, dx \, dy \tag{2.71}$$

where m, n are spatial frequencies in the x, y directions respectively, so that $1/m$, $1/n$ are the corresponding spatial wavelengths. We are thus considering the object as a superposition of gratings. The inverse transform relation gives us

$$t(x, y) = \int\int_{-\infty}^{+\infty} T(m, n) \exp -2\pi j (mx + ny) \, dm \, dn \tag{2.72}$$

and substituting this in equation (2.57) we have

$$U_3(x_3, y_3) = \frac{1}{\lambda^2 M d_1^2} \int\int\int_{-\infty}^{+\infty}\int T(m, n) h\left(x_1 + \frac{x_3}{M}, y_1 + \frac{y_3}{M}\right)$$

$$\times \exp -2\pi j (mx_1 + ny_1) \, dm \, dn \, dx_1 \, dy_1. \tag{2.73}$$

Performing the integrals in x_1, y_1 using the inverse of equation (2.50) we obtain

$$U_3(x_3, y_3) = \frac{1}{\lambda^2 M d_1^2} \int\!\!\!\int_{-\infty}^{+\infty} T(m, n) P(m\lambda d_1, n\lambda d_1)$$

$$\times \exp\frac{2\pi j}{M}(mx_3 + ny_3)\, dm\, dn. \qquad (2.74)$$

The image may thus be found by resolving the object transmittance into its Fourier spectrum, multiplying by a coherent transfer function, which gives the strength of the various Fourier components in the image, and then inverse transforming to give the image amplitude. We may thus write

$$U_3(x_3, y_3) = \frac{1}{\lambda^2 M d_1^2} \int\!\!\!\int_{-\infty}^{+\infty} T(m, n) c(m, n)$$

$$\times \exp\frac{2\pi j}{M}(mx_3 + ny_3)\, dm\, dn \qquad (2.75)$$

where the coherent transfer function $c(m, n)$ is given by

$$c(m, n) = P(m\lambda d_1, n\lambda d_1). \qquad (2.76)$$

The positive exponent in (2.75) indicates that the image is inverted. For the case of a circular pupil we may further write

$$c(m, n) = P(\tilde{m}a, \tilde{n}a) \qquad (2.77)$$

where $\tilde{m}$, $\tilde{n}$ are normalised (dimensionless) spatial frequencies and a is the lens radius.

If we now consider a line structure for which $n = 0$, then the transfer function is

$$\begin{aligned} c(m, 0) &= 1, & |\tilde{m}| &< 1, \\ &= 0, & |\tilde{m}| &> 1. \end{aligned} \qquad (2.78)$$

This spatial frequency cut-off corresponds to a spatial wavelength in the object of $v/2\pi$.

Let us consider an object which may be described by

$$t(x) = 1 + b\cos 2\pi v x, \qquad (2.79)$$

that is, it consists of only one pair of spatial frequency, one positive and one negative. The Fourier transform of the object is

$$T(m, n) = \left[\delta(m) + \frac{b}{2} \delta(m - v) + \frac{b}{2} \delta(m + v) \right] \delta(n). \qquad (2.80)$$

If the transfer function is even, we have for the image amplitude

$$U(x_3) = c(0) + bc(r) \cos 2\pi v x, \qquad (2.81)$$

and for the intensity

$$I(x_3) = |c(0)|^2 + \tfrac{1}{2}|b|^2|c(v)|^2 + 2 \, \mathrm{Re} \, \{c^*(0)c(v)b\} \cos 2\pi v x_3$$
$$+ \tfrac{1}{2}|b|^2|c(v)|^2 \cos 4\pi v x_3 \qquad (2.82)$$

where $\mathrm{Re} \, \{ \ \}$ denotes the real part and $*$ denotes the complex conjugate. If the modulus of b is small such that we can neglect $|b|^2$ this becomes

$$I(x_3) = |c(0)|^2 + 2 \, \mathrm{Re} \, \{c^*(0)c(v)b\} \cos 2\pi v x_3, \qquad (2.83)$$

which is a linear image of the object amplitude transmittance. If the transfer function is real, as it is for an aberration free pupil, and b is purely imaginary, there is no image.

For an object which has a cosinusoidal variation in absorption coefficient, refractive index or thickness (or height in a reflection specimen) we can write

$$t(x) = \exp \{b \cos 2\pi v x\}. \qquad (2.84)$$

In this case there are an infinite series of spatial frequencies present, but if $|b|$ is small we can expand as a power series

$$t(x) = 1 + b \cos 2\pi v x + \tfrac{1}{2}b^2 \cos^2 2\pi v x + \dots \qquad (2.85)$$

If we can neglect terms in higher order than b, equation (2.85) reduces to (2.79) and the image is given by equation (2.83). If the term of order b in equation (2.83) is zero, however, then we must include terms of order b^2 in equation (2.85), in which case equation (2.82) is no longer valid.

2.8   The angular spectrum

The amplitude of a plane wave of unit strength at a point $\mathbf{r}$ is given by

$$U(\mathbf{r}) = \exp -j(\mathbf{k} \cdot \mathbf{r}) \qquad (2.86)$$

where $\mathbf{k}$ is the wave vector such that if the direction cosines of the direction of propagation are α, β, γ, this may be written

$$U(x, y, z) = \exp \frac{-2\pi j}{\lambda} (\alpha x + \beta y + \gamma z), \qquad (2.87)$$

where α, β and γ are related by

$$\alpha^2 + \beta^2 + \gamma^2 = 1. \tag{2.88}$$

In the plane $z = 0$, the plane wave is thus

$$U(x, y, 0) = \exp \frac{-2\pi j}{\lambda} (\alpha x + \beta y). \tag{2.89}$$

If we illuminate an object with transmission $t(x, y)$ we have, in terms of the Fourier Spectrum of the object (equation 2.72),

$$t(x, y) = \int_{-\infty}^{+\infty} T(m, n) \exp -2\pi j (mx + ny) \, dm \, dn. \tag{2.90}$$

Comparing equations (2.89) and (2.90), we see that we can think of the amplitude immediately behind the object as being made up of many plane waves travelling in directions

$$\alpha = m\lambda, \qquad \beta = n\lambda, \tag{2.91}$$

where the strength of the particular plane wave is

$$T(m, n) = T(\alpha/\lambda, \beta/\lambda). \tag{2.92}$$

The limits of the integral in equation (2.90) require that α, β be allowed to vary in the range $-\infty$ to $+\infty$. Our assumptions of paraxial optics assume that α and β are small, and this condition is satisfied if the object transmittance is slowly varying relative to the wavelength. Otherwise waves with α and β greater than unity are produced. These evanescent waves decay quickly with z, as by equation (2.88) γ is complex. In our case the lens, assumed to be of small aperture, collects only waves for small α and β, and the presence of these other components need not concern us.

We now see a physical picture for the transfer function of an imaging system, for if a spectral component with spatial frequency m in the object results in a wave propagating at an angle θ to the optic axis we have from equation (2.90)

$$\theta \approx m\lambda \tag{2.93}$$

and the transfer function will cut off (Fig. 2.4) when

$$\theta \approx \frac{a}{d_1} \approx m\lambda \tag{2.94}$$

or

$$m = a/\lambda d_1 \qquad (2.95)$$

where a is the radius of the lens pupil. This is the basis of the Abbé theory of microscope imaging.

2.9   Incoherent imaging

In the previous sections of this chapter we have been considering coherent imaging systems. These systems produced images which were linear in amplitude, in the sense that the amplitude image of each point in the object transparency added to give the final amplitude image. The intensity image is given by the modulus square. We now consider the other extreme of incoherent imaging, which is linear in intensity such that the intensities of individual point images add. Such an object might be formed if it is self-luminous, if it emits light such that there is no phase coherence between the different points. Alternatively a transparency may be illuminated incoherently, as we shall discuss in the next chapter.

As the intensities in the images of the individual points add, we have for an object of amplitude transmittance $t(x_1, y_1)$

$$I(x_3, y_3) = \frac{1}{\lambda^4 M^2 d_1^4} \int\int_{-\infty}^{+\infty} |h|^2 \left(x_1 + \frac{x_3}{M}, y_1 + \frac{y_3}{M} \right) |t|^2(x_1 y_1) \, dx_1 \, dy_1$$

$$(2.96)$$

that is, the convolution of the intensity transmittance of the object and the intensity point spread function.

For a single point situated at $x_1 = x$, $y_1 = y$, the image intensity is given by

$$I(x_3, y_3) = \frac{1}{\lambda^4 M^2 d_1^4} \int\int_{-\infty}^{+\infty} |h|^2 \left(x_1 + \frac{x_3}{M}, y_1 + \frac{y_3}{M} \right) \delta(x_1 - x, y_1 - y) \, dx_1 \, dy_1$$

$$(2.97)$$

which leads to precisely the same result as equation (2.58) for the coherent case.

If we again consider line structures such that the transmittance is a function of one direction only, we can obtain from equation (2.95), by analogy to equation (2.62), an incoherent line spread function $g'(x_1)$, such that

$$g'(x) = \int_{-\infty}^{+\infty} |h|^2(x_1, y_1) \, dy_1. \qquad (2.98)$$

In general it is not possible to express this integral simply, in terms of the pupil function of the lens, as in the coherent case. However, by substituting the appropriate point spread function into equation (2.95), we obtain $g'(x_1)$ by direct integration. For a circular lens we have

$$g'(v) = \int_{-\infty}^{+\infty} \left[\frac{2J_1(v^2 + v'^2)}{(v^2 + v'^2)^{\frac{1}{2}}} \right]^2 dv' \qquad (2.99)$$

which may be written as [2.7]

$$g'(v) = \frac{3\pi}{8} \frac{H_1(2v)}{v^2} \qquad (2.100)$$

where H_1 is a first-order Struve function and where we have normalised to unity at the origin. The incoherent line spread function is illustrated in Fig. 2.7.

We may also consider the image of the straight edge (equation 2.68) which may be calculated from equation (2.100), for a system using circular lenses, as

$$I(v) = \frac{3\pi}{8} \int_{v}^{\infty} \frac{H_1(2z)}{z^2} dz \qquad (2.101)$$

$$\propto \frac{1}{\pi} \left\{ \frac{H_1(2v)}{2v} + \int_{2v}^{\infty} \frac{H_0(z)}{z} dz \right\}, \qquad (2.102)$$

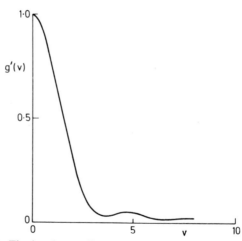

FIG. 2.7. The incoherent line spread function for a circular lens.

where H_0 is a zero order Struve function, and where we have used Struve's integral [2.7, p. 497] and normalised (2.102) to unity at large negative values of v. This is plotted in Fig. 2.6 and we see that the fringing which characterised the coherent response is absent. It is also important to note that the apparent position of the edge is different in the two cases. If we assume (arbitrarily) that the edge occurs at the position of the half-intensity response, we would introduce a slight error in the coherent case. We now finally consider incoherent imaging in terms of spatial frequencies. Following section 2.7 we introduce the object intensity spectrum $\tau(m, n)$, such that

$$|t|^2(x, y) = \int\limits_{-\infty}^{+\infty}\!\!\int \tau(m, n) \exp{-2\pi j(mx + ny)} \, dm \, dn. \qquad (2.103)$$

Substituting into equation (2.9), we have

$$I(x_3, y_3) = \frac{1}{\lambda^4 M^2 d_1^4} \int\limits_{-\infty}^{+\infty}\!\!\int\!\!\int\!\!\int |h|^2\!\left(x_1 + \frac{x_3}{M}, y_1 + \frac{y_3}{M}\right)$$

$$\times \tau(m, n) \exp{-2\pi j(mx_1 + ny_1)} \, dm \, dn \, dx_1 \, dy_1. \qquad (2.104)$$

Performing the integrals in x_1, y_1 and using the inverse of equation (2.50) together with the convolution theorem (see e.g. reference 2.1, p. 10), we may write

$$I(x_3, y_3) = \frac{1}{\lambda^4 M^2 d_1^4} \int\limits_{-\infty}^{+\infty}\!\!\int \tau(m, n) C(m, n) \exp{\frac{-2\pi j}{M}}(mx_3 + ny_3) \, dm \, dn,$$

$$(2.105)$$

with

$$C(m, n) = P(m\lambda d_1, n\lambda d_1) \otimes P^*(m\lambda d_1, n\lambda d_1) \qquad (2.106)$$

where $\otimes$ denotes the convolution operation.

This function may be called the incoherent transfer function. The similarity between equations (2.105) and (2.75) should be noted.

The incoherent transfer function is plotted in Fig. 2.8 for circular pupils. We see that it is non-zero for line structures with normalised spatial frequencies $(\tilde{m}, \tilde{n})$ less than *two*. This spatial frequency cut-off corresponds to a special wavelength in the object of v/π, that is *twice* the spatial frequency bandwidth of the coherent system. The smooth gradual fall off of this transfer function may be seen as the reason for the absence of ringing in the straight edge response.

The convolution of two circles is given by the area in common when the centre of one is displaced. For line structures the transfer function may be evaluated analytically to give

$$C(\tilde{m}) = \frac{2}{\pi}\left[\cos^{-1}\frac{\tilde{m}}{2} - \frac{\tilde{m}}{2}\left(1 - \left(\frac{\tilde{m}}{2}\right)^2\right)^{1/2}\right]. \tag{2.107}$$

We may also consider the images of weak objects, as in section 2.7. The remarks made there apply equally here, but with the proviso that $C(m, n)$ is now always *real*, even in the presence of aberrations, and thus only the real part of b is ever imaged.

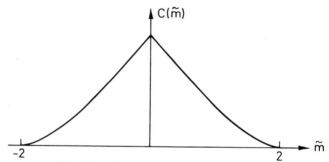

FIG. 2.8. The incoherent transfer function.

It would not be correct to give the impression that an incoherent system is always to be preferred. For example, we have just seen above that no phase information will ever be imaged. We should also be rather careful in comparing the cut-offs of the two transfer functions, as they are not strictly comparable: one is concerned with amplitudes and the other with intensities. However it is important to understand the differences between these two imaging modes. They can be regarded as two limits in microscope imaging which, as we shall see in the next chapter, is in general neither strictly coherent nor incoherent, but rather partially coherent.

References

[2.1] J. W. Goodman (1908). "Introduction to Fourier Optics". McGraw Hill, San Francisco.

[2.2] M. Born and E. Wolf (1975). "Principles of Optics". Pergamon Press, Oxford.

[2.3] G. B. Airy (1841). *Phil. Mag.* **18**, 1.

[2.4] G. C. Steward (1925). *Phil. Trans. R. Soc.* **A225**, 131.

[2.5] G. C. Steward (1928). "The Symmetrical Optical System". Cambridge University Press, Cambridge.

[2.6] W. T. Welford (1960). *J. Opt. Soc. Am.* **50**, 749.

[2.7] M. Abromowitz and I. A. Stegun (1965). "Handbook of Mathematical Functions". Dover, New York.

Chapter 3

Image Formation in Scanning Microscopes

We have already demonstrated, by physical arguments, the equivalence of the conventional microscope and the Type 1 scanning microscope and suggested the arrangement of the Type 2 or confocal scanning microscope which was predicted to have far superior imaging properties to conventional microscopes.

We now put these assertions on a more rigorous basis by using the Fourier imaging approach introduced in the previous chapter to compare the various imaging configurations.

3.1 Imaging with the STEM configuration

We begin our discussion of imaging in practical microscope systems by considering the arrangement of Fig. 3.1. Here we have one lens of pupil function $P_1(\xi_1, \eta_1)$ which focuses light onto the scanning object of amplitude transmittance $t(x_0, y_0)$; the transmitted radiation is then collected by a large area detector which has an amplitude detection sensitivity of $P_2(\xi_2, \eta_2)$. This is the same as the electron optical layout employed in the scanning transmission electron microscope and we should remember that although we are primarily concerned with light the remarks apply equally to electrons provided we assume appropriate functions for P_1, P_2 etc.

We can write the field just after passing through the object as

$$U_2(x_0, y_0; x_s, y_s) = h_1(x_0, y_0)t(x_0 - x_s, y_0 - y_s), \qquad (3.1)$$

where (x_s, y_s) represents the scan position and h_1 is the amplitude point spread function of the lens, given by (2.50)

$$h_1(x_0, y_0) = \int\int\limits_{-\infty}^{+\infty} P_1(\xi_1, \eta_1) \exp\frac{jk}{d}(\xi_1 x_0 + \eta_1 y_0)\, d\xi_1\, d\eta_1. \qquad (3.2)$$

The field at the detector may now be found by propagating $U_2(x_0, y_0; x_s, y_s)$ to the (ξ_2, η_2) plane using the Fraunhofer diffraction integral (2.5).

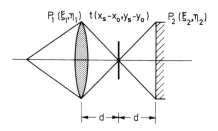

FIG. 3.1. The geometry of the STEM configuration.

Neglecting the premultiplying constants and phase variation this becomes

$$U_3(\xi_2, \eta_2; x_s, y_s) = \int\int\limits_{-\infty}^{+\infty} h_1(x_0, y_0) t(x_0 - x_s, y_0 - y_s)$$

$$\times \exp\frac{jk}{d}(\xi_2 x_0 + \eta_2 y_0)\, dx_0\, dy_0. \qquad (3.3)$$

The signal that is measured on the detector is

$$I(x_s, y_s) = \int\int\limits_{-\infty}^{+\infty} |U(\xi_2, \eta_2; x_s, y_s) P_2(\xi_2, \eta_2)|^2\, d\xi_2\, d\eta_2. \qquad (3.4)$$

Performing the integrals in (ξ_2, η_2) allows us to write

$$I(x_s, y_s) = \int\int\int\int\limits_{-\infty}^{+\infty} h_1(x_0, y_0) h_1^*(x_0', y_0') t(x_0 - x_s, y_0 - y_s)$$

$$\times t^*(x_0' - x_s, y_0' - y_s) g_2(x_0 - x_0', y_0 - y_0')\, dx_0\, dx_0'\, dy_0\, dy_0', \qquad (3.5)$$

where g is the spread function associated with the modulus squared of P_2, given by

$$g_2(x, y) = \int\int_{-\infty}^{+\infty} |P_2(\xi_2, \eta_2)|^2 \exp\frac{jk}{d}(x\xi_2 + y\eta_2)\,d\xi_2\,d\eta_2. \tag{3.6}$$

We now mention two limiting cases of this microscope. The first is when the detector is infinitely large and of constant sensitivity such that P_2 is always unity. Under these circumstances g_2 becomes a delta function and the image may be written, from equation (3.5), as

$$I(x_s, y_s) = |h_1|^2 \otimes |t|^2, \tag{3.7}$$

that is, the imaging is *incoherent*. On the other hand in the limit as P_2 becomes extremely small in extent the image reduces to

$$I(x_s, y_s) = |h_1 \otimes t|^2, \tag{3.8}$$

or *coherent* imaging. It is clear that for a finite sized detector the image may not be written in either of these simple forms and thus we must speak of the imaging being in general partially coherent with coherent and incoherent being the extreme limits.

We may gain further insight by considering the image of a few simple objects in such a microscope. A single point object, for example, has an image from equation (3.5)

$$I(x_s, y_s) = |h_1(x_s, y_s)|^2 g_2(0, 0), \tag{3.9}$$

and thus the image is independent of the sensitivity distribution and size of the detector, P_2. If we now move onto a two-point object where the points are separated by $(2x_d, 0)$ then

$$t(x_0, y_0) = \delta(y_0)[\delta(x_0 - x_d) + \delta(x_0 + x_d)]. \tag{3.10}$$

Now $|P_2|^2$ is real and if we assume it is even so that g_2 is also real and even, we may use equation (3.5) to write after renormalising

$$I(x_s, 0) = |h_1(x_s + x_d, 0)|^2 + |h_2(x_s - x_d, 0)|^2$$
$$+ \frac{2g_2(2x_d, 0)}{g_2(0, 0)} \text{Re}\,\{h_1(x_s + x_d, 0)h_1^*(x_s - x_d, 0)\}. \tag{3.11}$$

The first two terms may be recognised as the images of each point separately whilst the third represents an interference term given by the cross-product of the amplitude images of the two points. The relative strength of this interference term depends on $g_2(2x_d, 0)$, that is on the size of P_2. If we consider our two limiting cases we have for a vanishingly small detector

$$I(x_s, 0) = |h_1(x_s + x_d, 0) + h_1(x_s - x_d, 0)|^2, \qquad (3.12)$$

where the amplitude images add together and imaging is therefore coherent. Conversely for a large area detector

$$I(x_s, 0) = |h_1(x_s + x_d, 0)|^2 + |h_1(x_s + x_d, 0)|^2, \qquad (3.13)$$

and the intensity images add together, as one would expect for incoherent imaging. Generalising, we may thus say that altering the size of the detector alters the coherence of the system.

The question as to how close together the two points may come before they are said to be no longer resolved is not easy to answer as various criteria have been proposed giving different results. The two most widely used criteria are the Sparrow criterion which is concerned with the rate of change of the slope of the image at the midpoint and the Rayleigh criterion which somewhat arbitrarily states that the two points will be just resolved when the intensity at the midpoint is 0.735 times that at the points. The Rayleigh criterion was introduced for incoherent imaging with a circular aberation-free pupil, in which it corresponds to the condition that the first zero of the image of one point coincides with the position of the central peak of the image of the second point object. We will discuss two-point imaging a little further by supposing that we have two circular aberration-free pupils such that

$$P_1(r) = 1, \qquad r < a_1, \qquad (3.14)$$

$$P_2(r) = 1, \qquad r < a_2. \qquad (3.15)$$

We further introduce a parameter s, defined by

$$a_2 = sa_1. \qquad (3.16)$$

The value $s = 0$ corresponds to coherent imaging and $s \to \infty$ to incoherent imaging. Using these definitions we can immediately write from equation (3.6)

$$g_2(v) = \frac{2J_1(sv)}{sv}, \qquad (3.17)$$

where v is the normalised optical coordinate of equation (2.53). Equation (13.11) may now be rewritten as

$$I(v_s, 0) = \left(\frac{2J_1(v_s + v_d)}{v_s + v_d}\right)^2 + \left(\frac{2J_1(v_s - v_d)}{v_s - v_d}\right)^2$$
$$+ 2\left(\frac{2J_1(2sv_d)}{2sv_d}\right)\left(\frac{2J_1(v_s + v_d)}{v_s + v_d}\right)\left(\frac{2J_1(v_s - v_d)}{v_s - v_d}\right). \qquad (3.18)$$

Thus we see that the image depends on the size of the detector relative to that of the objective rather than the absolute size of the detector. The parameter s is called the coherence parameter. It is interesting to note that whenever $2sv_d$ is a root of $J_1(2sv_d) = 0$ the product term is absent and the image is the same as would have been obtained if the object had been incoherently illuminated. In particular for equal pupils ($s = 1$) this will be the case when $2v_d$ is a non-zero root of $J_1(2v_d) = 0$ which means practically that the geometrical images of the pinholes are separated by a distance equal to the radius of *any* dark ring of the Airy pattern of the objective. Thus if the two points are separated by a distance such that the Rayleigh criterion is satisfied for an incoherent system, the Rayleigh criterion is also satisfied for a system with equal objective and detector pupils. So the two-point resolution as given by the Rayleigh criterion is identical in these two systems.

We can discuss the effect of the coherence parameter s on the two-point resolution by introducing a function [3.1]

$$L(s) = 2v_d, \tag{3.19}$$

which is the distance in optical coordinates between two point objects such that the Rayleigh criterion

$$\frac{I(0,0)}{I(v_d, 0)} = 0.735 \tag{3.20}$$

is satisfied.

This function is plotted in Fig. 3.2 and we can see that the separation for the points to be just resolved for equal pupils or very large detector is 0.61 optical units. The best resolving power is obtained with $s \sim 1.5$. We have included in Fig. 3.2 the curve for a microscope employing a full circular

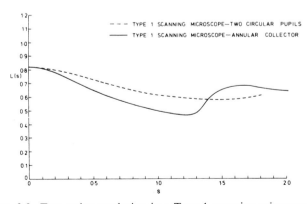

FIG. 3.2. Two point resolution in a Type 1 scanning microscope.

objective lens and an infinitely narrow annular detector. For the particular case of a two-point object the limiting resolution is improved by employing such a collector [3.2].

3.2   The partially coherent Type 1 scanning microscope

Although the arrangement of Fig. 3.1 is employed in the scanning transmission electron microscope it is usual in scanning optical microscopy to employ a second, collector lens to gather the radiation which has passed through the object and focus it onto the detector. Figure 3.3 shows two possible configurations in which the detector collects all the light which is incident on P_2 and so they both have exactly the same imaging properties as each other and as the STEM configuration. In Chapter 1 we discussed how the Type 1 scanning microscope is equivalent to the conventional microscope. In the scanning microscope of Fig. 3.3(a) the radiation is focused on to the detector, and as it is analogous to the critical illumination system in conventional microscopy it may be termed critical detection. In Fig. 3.3(b) on the other hand the detector is placed in the back focal plane of the collector lens and we may call this Köhler detection, again by analogy with Köhler illumination in conventional microscopy. The Köhler system relies on the response of the detector being uniform across the whole area and so the preferred approach is the critical detection arrangement of Fig. 3.3(a).

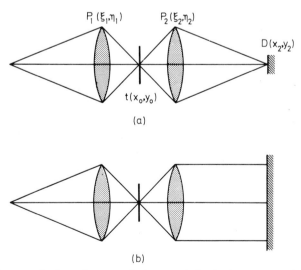

Fig. 3.3. Two equivalent forms of the partially coherent Type 1 scanning microscope.

This is in contrast to conventional microscopy where, as a complete object field must be imaged, Köhler illumination is preferred to give uniform illumination over the field.

The imaging of the Type 1 scanning microscope is still described by equations (3.5) and (3.6) but now $P_2(\xi_2, \eta_2)$ represents the pupil function of the collector lens if we assume that the detector has a uniform response. We now proceed to discuss the imaging in terms of spatial frequencies. This concept was introduced in section (2.7) where we represented a non-periodic object in terms of its Fourier transform or spectrum. Thus we can write (equation 2.72)

$$t(x, y) = \int\limits_{-\infty}^{+\infty}\!\!\int T(m, n) \exp\, -2\pi j (mx + ny) \, \mathrm{d}m \, \mathrm{d}n \qquad (3.21)$$

and for the complex conjugate

$$t^*(x, y) = \int\limits_{-\infty}^{+\infty}\!\!\int T^*(p, q) \exp\, 2\pi j (px + qy) \, \mathrm{d}p \, \mathrm{d}q \qquad (3.22)$$

where m, p are spatial frequencies in the x direction and n, q are similarly spatial frequencies in the y direction. We have to introduce the spatial frequencies p, q, which are dummy variables which disappear upon integration of (3.22), in order to be able to write the product of (3.21) and (3.22) with the integral signs at the front.

Substituting equations (3.21) and (3.22) into (3.5) we obtain

$$I(x_s, y_s) = \int\limits_{-\infty}^{+\infty}\!\!\int\!\int\!\int\!\int\!\int\!\int h_1(x_0, y_0) h_1^*(x_0', y_0') T(m, n) T^*(p, q)$$

$$\times\, g_2(x_0 - x_0', y_0 - y_0')$$

$$\times \exp\, -2\pi j \{ m(x_0 - x_s) - p(x_0' - x_s) + n(y_0 - y_s) - q(y_0' - y_s) \}$$

$$\times\, \mathrm{d}m \, \mathrm{d}n \, \mathrm{d}p \, \mathrm{d}q \, \mathrm{d}x_0 \, \mathrm{d}y_0 \, \mathrm{d}x_0' \, \mathrm{d}y_0'. \qquad (3.23)$$

This may be written as

$$I(x_s, y_s) = \int\limits_{-\infty}^{+\infty}\!\!\int\!\int\!\int C(m, n; p, q) T(m, n) T^*(p, q)$$

$$\times \exp\, 2\pi j \{ (m - p)x_s + (n - q)y_s \} \, \mathrm{d}m \, \mathrm{d}n \, \mathrm{d}p \, \mathrm{d}q, \qquad (3.24)$$

with

$$C(m, n; p, q) = \int\int\int\int_{-\infty}^{+\infty} h_1(x_0, y_0) h_1^*(x_0', y_0') g_2(x_0 - x_0', y_0 - y_0')$$

$$\times \exp{-2\pi j \{m x_0 - p x_0' + n y_0 - q y_0'\}} \, dx_0 \, dy_0 \, dx_0' \, dy_0'$$

(3.25)

which using equations (3.6) and (3.2) may be recast as

$$C(m, n; p, q) = \int\int_{-\infty}^{+\infty} |P_2(\xi_2, \eta_2)|^2 P_1(\xi_2 + m\lambda d, \eta_2 + n\lambda d)$$

$$\times P_1^*(\xi_2 + p\lambda d, \eta_2 + q\lambda d) \, d\xi_2 \, d\eta_2.$$

(3.26)

This represents a very important result as we have been able to express the intensity variation in the image of an arbitrary specimen by equation (3.24) in which $C(m, n; p, q)$, the partially coherent transfer function (sometimes also called the transmission cross coefficient), is a function *only* of the optical system and *not* the object. This is the real power of this approach whereby we can introduce an imaging function which is common to all objects. We can see from equation (3.24) that "perfect" imaging is obtained if the transfer function $C(m, n; p, q)$ is always unity; this is not possible in practice and the aim in microscope design is to make this function as smooth and great in extent as possible. It should be noted, however, that a "perfect" image does now show up phase variations, so that "perfect" imaging may not even be desirable in practice.

In order to fix our ideas concerning the transfer function method let us now consider a line structure which has detail in the x-direction only. The transfer function may now be contracted to

$$C(m; p) = C(m, 0; p, 0)$$

(3.27)

$$= \int\int_{-\infty}^{+\infty} |P_2(\xi_2, \eta_2)|^2 P_1(\xi_2 + m\lambda d, \eta_2) P_1^*(\xi_2 + p\lambda d, \eta_2) \, d\xi_2 \, d\eta_2,$$

(3.28)

and $C(m; p)$ is the transfer function which gives the magnitude of the spatial frequency component $(m - p)$ in the *intensity* image.

We further consider the case where the microscope has aberration-free circular pupils of the form of equations (3.14) and (3.15). We may graphically calculate the transfer function as the area of overlap of the two circles representing P_1 centred on $(-m\lambda d, 0)$ and $(-p\lambda d, 0)$, which also falls within the circle P_2. This is illustrated in Fig. 3.4. There are two limiting cases of interest as the diameter of P_2 is varied. When P_2 is large we see that $C(m; p)$ becomes a function of $(m - p)$ only. This is in fact what one expects for incoherent imaging, and imaging is indeed incoherent for this limiting case as

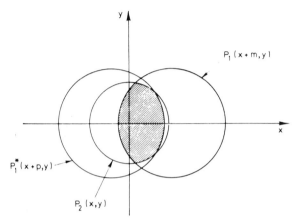

FIG. 3.4. The region of integration for $C(m; p)$.

discussed in section (2.9). On the other hand as P_2 becomes vanishingly small we find that

$$C(m; p) = P_1(m\lambda d)P_1^*(p\lambda d) \qquad (3.29)$$

$$= c(m)c^*(p), \qquad (3.30)$$

and thus the image may be written as

$$I(x_s) = \left| \int_{-\infty}^{+\infty} c(m)T(m) \exp (2\pi jmx_s)\, dm \right|^2, \qquad (3.31)$$

which of course corresponds to the coherent imaging of section (2.7). Again the positive exponent corresponds to an inverted image.

A practically important case is where the two pupils are of equal size. Under these circumstances the imaging is partially coherent and the $C(m; p)$ function somewhat more complicated. It is shown in Fig. 3.5(b) in $(m; p)$ space where it is seen to exhibit an hexagonal cut-off. It should be remembered that although m and p are plotted here in orthogonal directions they represent two spatial frequencies in the *same* direction. The symmetry of the surface is also shown, $\Lambda(\tilde{v})$ being the radial variation of the convolution of two circles which may be written

$$\Lambda(\tilde{v}) = \frac{2}{\pi}\left[\cos^{-1}\left(\frac{\tilde{v}}{2}\right) - \left(\frac{\tilde{v}}{2}\right)\left\{1 - \left(\frac{\tilde{v}}{2}\right)^2\right\}^{1/2}\right], \quad |\tilde{v}| < 2. \quad (3.32)$$

Here $\tilde{v}$ is the normalised spatial frequency given by $v\lambda d/a$, where a is the radius of the pupil, so that the cut off is at $\tilde{v} = 2$.

We also show, for comparison, the transfer functions for the coherent and incoherent microscopes in Fig. 3.5. It is clear that Fig. 3.5(b) represents a transition between the two extremes. The $C(m; p)$ surfaces are shown in Fig. 3.6.

The Type 1 scanning microscope behaves in an identical way to the analogous conventional microscope. Imaging in partially coherent conventional microscopes was analysed by Hopkins [3.3] using the theory of partial coherence. The spread function g_2 of equation (3.18) may be recognised as nothing other than the degree of coherence [3.4] which may be derived from the van Cittert–Zernike theorem [3.5, p. 510].

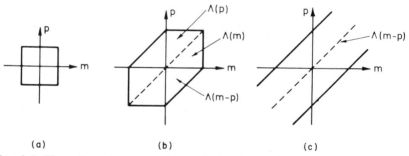

FIG. 3.5. The regions of non-zero $C(m; p)$ in (m, p) space for (a) coherent, (b) partially coherent and (c) incoherent microscope.

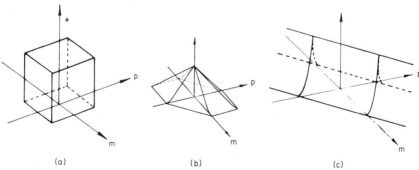

FIG. 3.6. The $C(m; p)$ surfaces for (a) coherent, (b) partially coherent and (c) incoherent microscopes.

3.3   The confocal scanning microscope

3.3.1   *Introduction*

As explained in section 1, the Type 1 scanning microscope has imaging properties identical to those of conventional non-scanning microscopes. The Type 2, or confocal scanning microscope, on the other hand has completely different imaging properties. The confocal microscope is formed by placing a point detector in the detector plane of Fig. 3.3(a). We can write the field in the detector plane (x_2, y_2) as the convolution of the amplitude in the object plane with the point spread function of the collector lens

$$U(x_2, y_2; x_s, y_s)$$

$$= \int\int_{-\infty}^{+\infty} h_1(x_0, y_0)t(x_0 - x_s, y_0 - y_s)h_2\left(\frac{x_2}{M} - x_0, \frac{y_2}{M} - y_0\right)dx_0\,dy_0.$$

$$(3.33)$$

However if we employ a point detector at $x_2 = y_2 = 0$ the detected intensity is

$$I(x_s, y_s) = \left|\int\int_{-\infty}^{+\infty} h_1(x_0, y_0)t(x_0 - x_s, y_0 - y_s)h_2(-x_0, -y_0)\,dx_0\,dy_0\right|^2,$$

$$(3.34)$$

which may be written for even spread functions

$$I(x_s, y_s) = |h_1 h_2 \otimes t|^2, \qquad (3.35)$$

that is the microscope behaves as a coherent microscope with an effective point spread function given by the product of those for the two lenses.

We have thus found that the combination of point detector and scanning converts a convolution of the form $(h_1 t) \otimes h_2$ (which is obtained for a Type 1 scanning microscope (equation 3.16)) into one of the form $(h_1 h_2) \otimes t$. Another way of considering the confocal microscope is to calculate first the amplitude U in terms of U_2,

$$U(x_2, y_2) = \int\int_{-\infty}^{+\infty} U_2(\xi_2, \eta_2)\exp\frac{jk}{Md}(\xi_2 x_2 + \eta_2 y_2)\,d\xi_2\,d\eta_2 \qquad (3.36)$$

so that if $x_2 = y_2 = 0$

$$I(0, 0) = \left|\int\int_{-\infty}^{+\infty} U_2(\xi_2, \eta_2)\,d\xi_2\,d\eta_2\right|^2, \qquad (3.37)$$

that is the effect of the point detector is to integrate the *amplitude* over the pupil P_2. This compares with the Type 1 scanning microscope where the detector integrates the *intensity* over the pupil P_2. Substituting for U_3 from (3.3) we obtain for the confocal case

$$I(x_s, y_s) = \left| \int\int\int\int_{-\infty}^{+\infty} h_1(x_0, y_0)t(x_0 - x_s, y_0 - y_s)P_2(\xi_2, \eta_2) \right.$$

$$\left. \times \exp\frac{jk}{d}(x_0\xi_2 + y_0\eta_2)\,\mathrm{d}x_0\,\mathrm{d}y_0\,\mathrm{d}\xi_2\,\mathrm{d}\eta_2 \right|^2. \tag{3.38}$$

and using (3.2) to introduce h_2 we reproduce (3.34).

The point detector integrates amplitude over the pupil P_2: it therefore has the same effect as the amplitude-sensitive detector used in acoustic microscopy [3.6]. The scanning acoustic microscope is a confocal microscope and exhibits many of the properties of confocal microscopes.

3.3.2   *Image formation in confocal microscopes*

If the two lenses in a confocal microscope are circular and of equal numerical aperture the image of a point object is (from 3.35)

$$I(v) = \left(\frac{2J_1(v)}{v}\right)^4, \tag{3.39}$$

which is shown in Fig. 3.7, the central peak being sharpened up by 27% relative to the image in a conventional microscope (at half the peak intensity). The sidelobes are also dratically reduced, so there is thus a marked reduction in the presence of artefacts in confocal images.

If we calculate the image of two closely spaced point objects we find that when the Rayleigh criterion is satisfied the points are separated by a normalised distance $2v_d = 0.56$. This is 32% closer than in a conventional coherent microscope and 8% closer than in a conventional microscope with equal lens pupils. The relative values are illustrated in Fig. 3.2. The fact that the sidelobes are weaker in confocal microscopy suggests that it should be possible to use an annular lens in a confocal microscope. With one circular and one annular lens of equal radii the intensity is given by

$$I(v) = \left(\frac{2J_0(v)J_1(v)}{v}\right)^2 \tag{3.40}$$

so that the central peak is now even narrower (40% narrower compared with a conventional microscope) and the sidelobes are extremely weak as the maxima of the spread function of the annulus coincide with the zeros of that of the circular lens. The two-point resolution is now 28% better than a conventional microscope with equal lens pupils.

For large values of v the intensity in the image of a single point in a conventional microscope falls off as v^{-3}. In a confocal microscope with two circular lenses it falls off as v^{-6}, whereas with one annular lens it falls off as v^{-4}. With two narrow annuli however it falls off as v^{-2}, that is the power in successive sidelobes only falls off as v^{-1} and total normalised power does not converge, so that such an arrangement is clearly unusable.

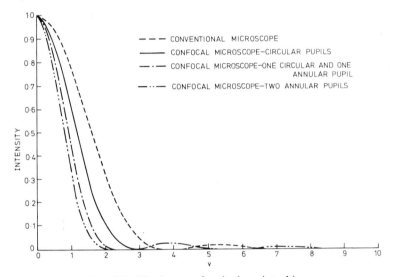

FIG. 3.7. The image of a single point object.

We may now turn our attention to the Fourier imaging and substituting equation (3.21) into (3.34) are able to write

$$I(x_s, y_s) = \left| \int\!\!\int_{-\infty}^{+\infty} c(m, n)T(m, n) \exp 2\pi j(mx_s + ny_s) \, dm \, dn \right|^2, \quad (3.41)$$

with

$$c(m, n) = P_1(m\lambda d, n\lambda d) \otimes P_2(m\lambda d, n\lambda d), \quad (3.42)$$

where $c(m, n)$ is a coherent transfer function. For two circular pupils of equal radii the *coherent* transfer function is identical to the *incoherent* transfer function for an incoherent system (equation 2.106).

If we again restrict ourselves to considering the images of line structures the function of interest is $C(m; p)$ given by

$$C(m; p) = c(m)c^*(p) \tag{3.43}$$

for the confocal microscope.

Figures 3.8(a) and 3.9(a) show the form of this transfer function for a confocal microscope with equal pupils. For a given pair of spatial frequency moduli the response is higher if they have the same sign (difference frequencies) than if of opposite sign (sum frequencies). The m–p axis is shown in Figs 3.8(a) and 3.8(b): the greater the distance along the axis the higher

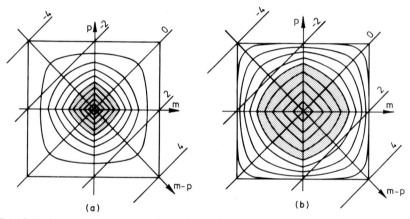

FIG. 3.8. Contours of constant $C(m; p)$ showing lines of normalised spatial frequency $(m - p)$ for (a) circular lenses and (b) one annular and one circular lens in confocal microscopes.

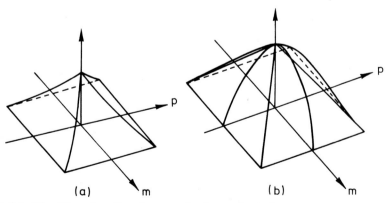

FIG. 3.9. The $C(m; p)$ surface for a confocal scanning microscope with (a) circular lenses and (b) with one annular lens and one circular lens.

the spatial frequency in the image. For the confocal microscope the response
for the sum frequencies is improved, but for the difference frequencies is
reduced as compared to the conventional microscope, Fig. 3.5(b). This
accounts for the fact that the imaging in confocal microscopy is generally
improved even though the *coherent* transfer function in the confocal microscope is identical to the *incoherent* transfer function for a conventional
incoherent microscope. This coherent transfer function is compared with that
for a conventional coherent microscopy in Fig. 3.10, the cut-off frequency
being twice as great. The transfer function also falls off gradually and thus we
do not expect excessive fringing to be present in the image of a straight edge.
Also shown is the transfer function for a confocal microscope with one
annular pupil, illustrating that the response for higher spatial frequencies is
improved. This transfer function is given by the radial variation of the
convolution of a circle with an annulus, given by

$$A(\tilde{v}) = \frac{2}{\pi} \cos^{-1}\left(\frac{\tilde{v}}{2}\right), \qquad \tilde{v} < 2. \qquad (3.44)$$

This should be compared with equation (3.32) for the convolution of two
circles.

The confocal microscope with one annular pupil may be compared with
that with two circular pupils by studying the region of m, p space within

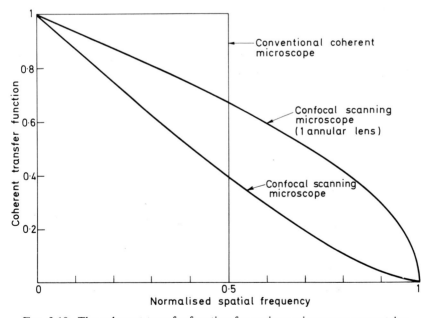

FIG. 3.10. The coherent transfer function for various microscope geometries.

which the transfer function is greater than one half, as illustrated by shading in Fig. 3.8.

If we now consider the imaging of a weak object of the form of equation (2.84), that is

$$t(x) = \exp(b \cos 2\pi vx) \tag{3.45}$$

with b small so that terms in b^2 may be neglected, we find that by substituting in equation (3.24) the image is given by

$$I(x_s) = C(0;0) + 2\,\mathrm{Re}\,\{bC(v;0)\}\cos 2\pi vx_s, \tag{3.46}$$

that is it depends only on $C(v;0)$. Imaging of weak objects in conventional and confocal microscopes with circular pupils is thus identical. However, if aberrations are present this is no longer the case and the confocal microscope may behave very differently.

3.4   Aberrations in scanning microscopes

3.4.1   *Introduction*

Until now we have discussed the imaging properties of various microscopes assuming perfect lenses. In practice the lenses will not be perfect and it is important to know how seriously lens aberrations or defocus affect the optical performance of the microscope. The most common practical arrangement of scanning optical microscope that has been constructed involves mechanically scanning the object across a stationary spot. This has the advantage that the optical system is axial and so the lenses need only strictly be corrected for axial aberrations, although in practice high-quality microscope objectives should be used if specially corrected lenses are not available.

In this section we therefore restrict our attention to the axial aberrations and consider the effects of defocus and primary spherical aberration. Chromatic aberration is not considered specifically as usually only a single laser wavelength is used. These effects are studied not only to examine the degradation of the imaging but also because in certain circumstances they serve to introduce an imaginary part to the transfer function and advantage could be taken of this to crudely image phase detail. We can see from equations (3.45) and (3.46) that if b is complex the intensity may be written as

$$I(x_s) = 1 + 2(b_r C_r - b_i C_i)\cos 2\pi vx_s, \tag{3.47}$$

where we have taken $C(0;0)$ to be unity and set

$$b = b_r + jb_i \tag{3.48}$$

and

$$C(v;0) = C_r + jC_i. \tag{3.49}$$

3.4.2 Defocus in scanning microscopes

We begin by considering the conventional or scanning microscope of Type 1; we see from equation (3.28) that for equal sized pupils the function $C(m; 0)$ is wholly real. Furthermore, as it is given by the convolution of a function with its complex conjugate, the aberrations of the collector lens P_2 are unimportant, this being equivalent to the well-known result in conventional microscopy that the aberrations of the condenser do not affect the imaging.

We will restrict ourselves to one-dimensional pupils for ease of analysis and introduce aberrated pupil functions, following equations (2.35), (2.37),

$$\left.\begin{aligned}P_1(x) &= \exp\left\{\tfrac{1}{2}ju_1\left(\frac{x}{a_1}\right)^2\right\}, \qquad |x| < a_1, \\ P_2(x) &= \exp\left\{\tfrac{1}{2}ju_2\left(\frac{x}{a_1}\right)^2\right\}, \qquad |x| < a_2,\end{aligned}\right\} \tag{3.50}$$

where

$$\left.\begin{aligned}u_1 &= 4k\,\delta z_1\sin^2(\alpha_1/2), \\ u_2 &= 4k\,\delta z_2\sin^2(\alpha_1/2).\end{aligned}\right\} \tag{3.51}$$

The wholly real transfer function for the conventional microscope is shown in Fig. 3.11 for varying degrees of defocus. The similarity of this result with

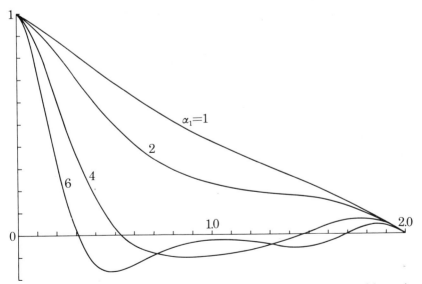

FIG. 3.11. The transfer function $C(m; 0)$ for a conventional microscope with varying degrees of defocus.

that obtained by Hopkins [3.7] for full circular pupils justifies our one-dimensional model. An imaginary part may be introduced into the transfer function by stopping down the collector lens. It is found, however [3.8], that this lens must be stopped down considerably before phase imaging becomes appreciable which, with the associated reduction in spatial frequency cut-off, is the major reason why this method of obtaining phase contrast is not widely used.

The properties of the confocal microscope, however, are very different. We recall that (equations 3.42, 3.43)

$$C(m; 0) = \{ P_1(\tilde{m}a) \otimes P_2(\tilde{m}a) \} \{ P_1^*(0) \otimes P_2^*(0) \}, \qquad (3.52)$$

$\tilde{m}$ again representing the normalised spatial frequency.

If the defocus of the pupils is equal and opposite, such that $P_1 = P_2^*$ the

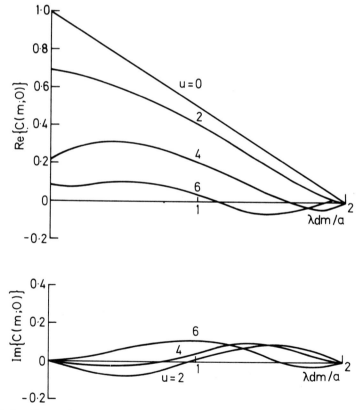

FIG. 3.12. The real and imaginary parts of the coherent transfer function of a transmission mode confocal microscope with lens defocus, or a reflection mode confocal microscope.

transfer function for weak objects is always real. Furthermore it is of exactly the same form as for the conventional microscope in Fig. 3.11. In a transmission instrument this corresponds to the case when the object is moved along the axis relative to the stationary lenses. For an object of uniform optical thickness, therefore, if the lenses are correctly spaced, no phase imaging will result from any depth within the object. Conversely if the pupils are different, for example one circular and one annular, then phase imaging will result from the defocused parts.

We now consider the case of equal defocus, $P_1 = P_2$, which corresponds to a displacement of both lenses such that the object remains midway between them. This is always the case in a reflection microscope and is the result of moving either the lens or the object: an imaginary part is introduced for the defocused parts of an object with variations in height. Equal defocus is also introduced if a transmission object has varying optical thickness which alters

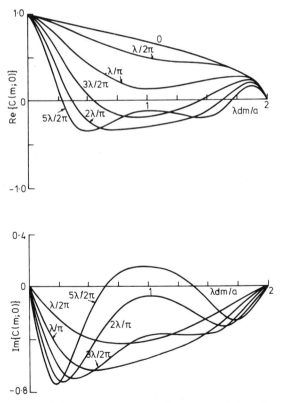

FIG. 3.13. The real and imaginary parts of the coherent transfer function of a confocal microscope with one annular pupil with varying degrees of defocus.

the effective spacing of the lenses. The transfer function is shown in Fig. 3.12 and we see that the imaginary part experiences a sign change which effectively rules it out as a practical method of phase contrast imaging. Care should therefore be taken when focusing by maximising the contrast if the object has comparitively strong phase variations. The strength of the transfer function at zero and low spatial frequencies is reduced with defocus by the spreading of the radiation in the detector plane. This property is associated with depth discrimination, which is discussed in section 3.7.

If we use an infinitely thin annular lens in a confocal microscope defocus of either object or the lens has the same result due to the large depth of focus of an annular lens. The transfer function is again complex (Fig. 3.13), a very small wavelength aberration ($\sim \lambda/2\pi$) being necessary to produce a substantial imaginary part to the transfer function. This is a major disadvantage of using an annular lens in a confocal microscope as parts of a thick object away from the focal plane are imaged with poor fidelity.

3.4.3   Defocus and primary spherical aberration in scanning microscopy

We now consider pupil function of the form

$$P(\xi) = \exp j\phi(\xi), \qquad |\xi| < 1, \tag{3.53}$$

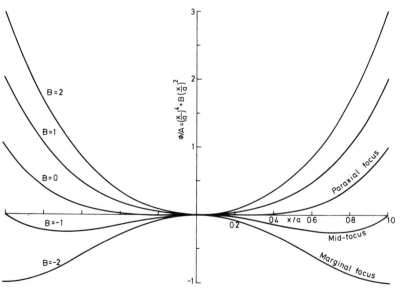

FIG. 3.14. The wave distortion function in the presence of spherical aberration and defocus.

where the wave distortion function is given by

$$\phi(\xi) = A\{(\xi/a)^4 + B(\xi/a)^2\}. \tag{3.54}$$

The coefficient A measures the total amount of fourth-power error on the wavefronts in the clear aperture, whereas the coefficient B specifies the focal setting, $B = 0$ corresponding to the paraxial setting, $B = -2$ to the marginal one while $B = -1$ indicates a "mid-focus" setting. For this last condition the wave deviations vanish at the centre and edge of the clear aperture (Fig. 3.14).

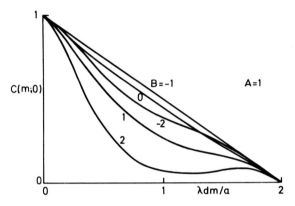

FIG. 3.15. The transfer function $C(m; 0)$ for a Type 1 scanning microscope with equal pupils for $A = 1$.

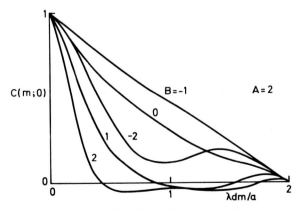

FIG. 3.16. The transfer function $C(m; 0)$ for a Type 1 scanning microscope with equal pupils for $A = 2$.

It should be noted that, unlike the case of pure defocus, the system is not symmetrical in focal setting and that by making B negative the effects of the spherical aberration may be to a certain extent be alleviated. We now consider the effect of this pupil function on the performance of various microscope types.

For the conventional microscope, $C(m; 0)$ is wholly real and independent of the aberrations of the second lens. This has been evaluated numerically and is plotted for various focal settings with degrees of sperical aberration

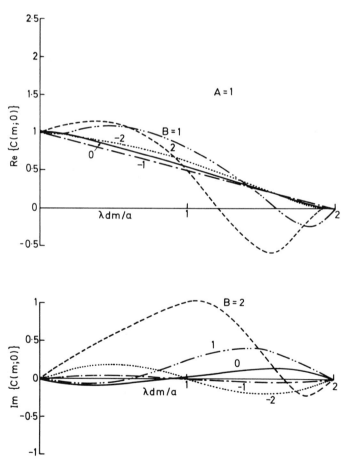

FIG. 3.17. The real and imaginary parts of the transfer function $C(m; 0)$ for a confocal scanning microscope with equal lenses, $A = 1$. Defoci of both lenses are equal.

corresponding to $A = 1$ and 2 in Figs 3.15 and 3.16 again for a one-dimensional model. The effect of the spherical aberration is drastically to reduce the mid- to higher spatial frequency components, but the effect is almost focused out at $B = -1$, the mid-focus setting. Positive values of B, on the other hand, reduce the performance and with higher degrees of spherical aberration and defocus phase reversal occurs for some spatial frequencies.

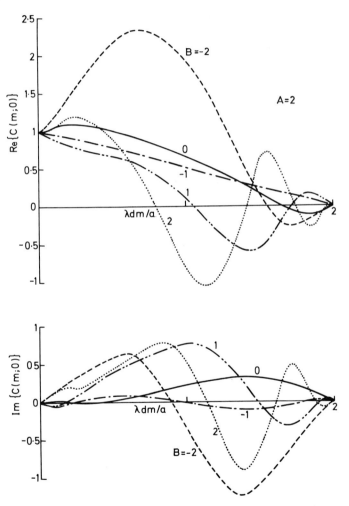

FIG. 3.18. The real and imaginary parts of the transfer function $C(m; 0)$ for a confocal scanning optical microscope with equal lenses, $A = 2$. Defoci of both lenses are equal.

We now turn to confocal scanning microscopes [3.9]. Here the aberrations of both lenses are important and the transfer function is in general complex. It is reasonable to assume that if two equal lenses are used that they would both suffer from similar degrees of spherical aberration and we have assumed that the coefficients A for the lenses are equal. We have plotted the transfer function for two special cases. The first is when the lenses have equal degrees of defocus (corresponding to a change in the separation of the lenses). It is seen (Figs 3.17 and 3.18) that again the mid-focus setting almost cancels out the effect of spherical aberration.

The second special case is that of equal and opposite defocus of the two lenses (corresponding to a movement of the specimen relative to the fixed lenses). The curves (Figs 3.19 and 3.20) indicate that defocus does not improve the high spatial frequency response and does not decrease the magnitude of the imaginary part. The curves for positive and negative defocus are identical because of the commutative property of the convolution operation.

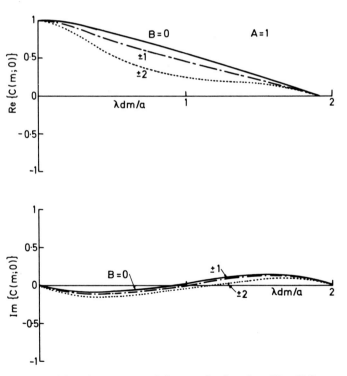

FIG. 3.19. Real and imaginary parts of the transfer function $C(m; 0)$ for a confocal scanning microscope with equal pupils, $A = 1$. Defocus of one lens is equal and opposite to that of the other.

The results of using an annular objective in a confocal microscope [3.10] are shown in Figs 3.21 and 3.22. These are equally applicable to conventional microscopes with an annular condenser. There is again no "aberrated" contribution from the annulus and so the curves apply equally to reflection microscopy. We can see that for $A = 1$ and 2 the mid-focus setting, $B = -1$ has almost cancelled out the effect of spherical aberration. However at higher values of A it is more difficult to focus out these effects and so a confused image consisting of both amplitude and phase information would result.

3.4.4 Discussion

In the conventional or Type 1 scanning microscope with two equal pupils the transfer function is purely real and although spherical aberration results in a degradation of the spatial frequency response the effect may be reduced by appropriate defocusing.

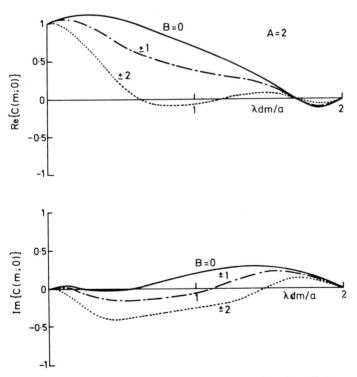

FIG. 3.20. Real and imaginary parts of the transfer function $C(m; 0)$ for a confocal scanning microscope with equal pupils, $A = 2$. Defocus of one lens is equal and opposite to that of the other.

In a confocal microscope, there is an imaginary part introduced. This may be reduced by adjustment of the lens spacing but not by movement of the object relative to the lenses. If the object transmittance exhibits both phase and amplitude variations interpretation of the micrographs may prove difficult. Thus, in general, for a confocal microscope to operate properly it is important that the lenses should be well corrected for the laser wavelength used, but, on the other hand, if the object is mechanically scanned off-axis aberrations are of course, unimportant. If the object is such that it exhibits only very small amplitude variations in transmittance the spherical aberration could result in useful phase imaging without this being associated with the detrimental reduction in spatial frequency bandwidth introduced by stopping down the collector lens.

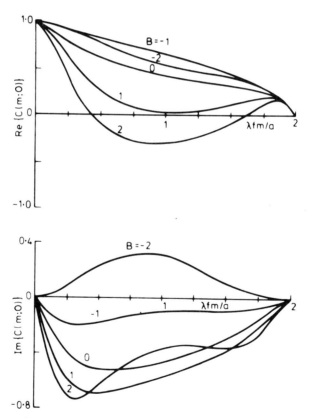

FIG. 3.21. Transfer function $C(m; 0)$ for a conventional microscope with annular condenser or confocal microscope with one annular lens in the presence of spherical aberration, $A = 1$.

The scanning acoustic microscope has imaging properties similar to those of the confocal optical microscope and the imaging is characterised by a coherent transfer function identical to the $C(m; 0)$ function of the confocal microscope. A typical figure for the maximum path error in an acoustic microscope might be $\lambda/4$. A value of A of unity corresponds to a maximum path error of $\lambda/6\cdot3$ and thus the spherical aberration in the acoustic microscope could well give rise to phase imaging.

We should note that although in some cases the effect of aberrations on the real part may be small care should be taken in the interpretation of micrographs especially if the phase delay in the object is appreciable.

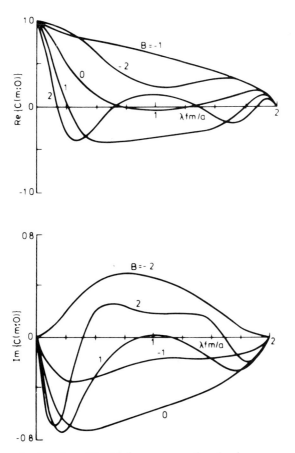

FIG. 3.22. Transfer function $C(m; 0)$ for a conventional microscope with annular condenser or confocal microscope with one annular lens in the presence of spherical aberration, $A = 2$.

3.5 The image of a straight edge

In sections 2.6 and 2.9 we calculated the image of a straight edge in coherent or incoherent imaging systems by finding the response to a δ-line and integrating. This method is not the most convenient for partially coherent systems. Instead we use the object spectrum, and in order that the integrals reduce to summations choose to consider the response to a bar pattern object and arrange for its spatial frequency to be sufficiently small that neighbouring edges do not affect each other [3.11]. We consider an object of the form (Fig. 3.23)

$$t(x) = M - \frac{2A}{\pi} \sum_{n=1}^{\infty} \frac{(-)^n}{2n-1} \cos(2n-1)2\pi vx, \qquad (3.55)$$

which when Fourier transformed and substituted in equation (3.24) gives

$$I(x) = M^2 - \frac{4AM}{\pi} \sum_{n=1}^{\infty} \frac{(-)^n}{2n-1} C\{(2n-1)v;0\} \cos(2n-1)\theta$$

$$+ \frac{2A^2}{\pi^2} \sum_{n=1}^{\infty} \sum_{r=1}^{\infty} \frac{(-)^{r+n}}{(2r-1)(2n-1)}$$

$$\times [C\{(2n-1)v, (2r-1)v\} \cos\{(2n-1)\theta - (2r-1)\theta\}$$

$$+ C\{(2n-1)v, -(2r-1)v\} \cos\{(2n-1)\theta + (2r-1)\theta\}], \qquad (3.56)$$

where

$$\theta = 2\pi vx, \qquad (3.57)$$

and we have used the identities

$$C(m;p) = C(-m;-p), \qquad (3.58)$$

$$C(m;0) = C(0;m). \qquad (3.59)$$

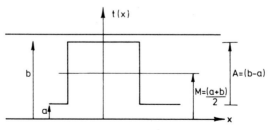

FIG. 3.23. The periodic object used in the calculation of the image of a straight edge object.

For the conventional microscope, the transfer function of which is shown in Fig. 3.24, we may write

$$I(x) = M^2 - \frac{4AM}{\pi} S_1 + \frac{2A^2}{\pi^2} S_2 + \frac{2A^2}{\pi^2} S_3 + \frac{4A^2}{\pi^2} S_4 + \frac{4A^2}{\pi^2} S_5, \quad (3.60)$$

where the Ss represent sums over portions of the $C(m; p)$ surface as shown in Fig. 3.2 and are given by

$$
\left.
\begin{aligned}
S_1 &= \sum_{n=1}^{\infty} \frac{(-)^n}{(2n-1)} C\{(2n-1)v\} \cos (2n-1)\theta, \\
S_2 &= \sum_{n=1}^{\infty} \frac{C\{(2n-1)v\}}{(2n-1)^2}, \\
S_3 &= \sum_{n=1}^{\infty} \frac{C\{2(2n-1)v\}}{(2n-1)^2} \cos \{2(2n-1)\theta\}, \\
S_4 &= \sum_{n=1}^{\infty} \sum_{r=n+1}^{\infty} \frac{(-)^{r+n} C\{(2r-1)v\}}{(2r-1)(2n-1)} \cos \{(2n-1)\theta - (2r-1)\theta\}, \\
S_5 &= \sum_{n=1}^{\infty} \sum_{r=n+1}^{\infty} \frac{(-)^{r+n} C\{(2n-1)v + (2r-1)v\}}{(2r-1)(2n-1)} \\
&\quad \times \cos \{(2n-1)\theta + (2r-1)\theta\},
\end{aligned}
\right\} \quad (3.61)
$$

where $C(\ )$ is either the convolution of two circular pupils for which equation (3.32) holds, or the convolution of a circle with a thin annulus (equation 3.44). The straight edge responses for conventional microscopes with two circular pupils and with an annular condenser and circular objective are

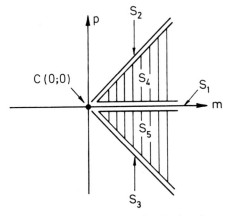

FIG. 3.24. The area of summation in (m, p) space.

shown in Fig. 3.25 where the normalised distance v is given by equation (2.28). It should be noted that the image in the microscope with an annular condenser is relatively poor. Turning now to the confocal microscope we obtain the much simpler result

$$I(x) = \left[M - \frac{2A}{\pi} \sum_{n=1}^{\infty} \frac{(-)^n C\{(2n-1)v\}}{(2n-1)} \cos\{(2n-1)\theta\} \right]^2. \qquad (3.62)$$

The relevant responses are plotted in Fig. 3.25 and we see that the confocal microscope with two circular lenses gives a better image than a conventional microscope but that a confocal microscope with one annular and one full circular lens gives the best response of all.

In a confocal microscope the use of two equal pupils results in an amplitude point spread function which is never negative and hence the straight edge response under these conditions does not exhibit fringing even

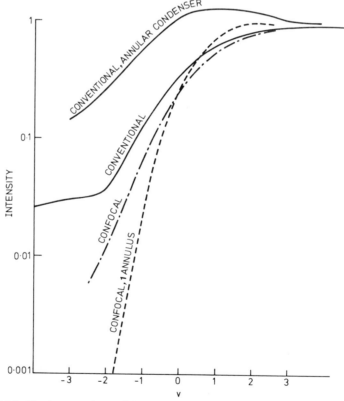

FIG. 3.25. The images of a straight edge object in various microscope arrangements.

when the object is defocused. There are thus advantages in using two equal pupils in a confocal microscope and we conclude this section by examining this case.

The impulse response for a confocal microscope with circular lenses has very weak side lobes. We know that the use of annular lenses results in a sharper central peak and also an increase in the strength of the side lobes. It is

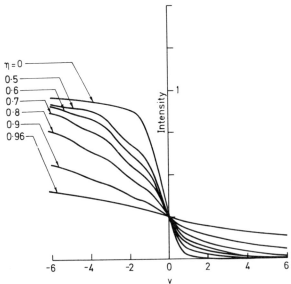

FIG. 3.26. The straight edge image in a confocal microscope with equal annular lenses.

interesting to examine which of these two effects is dominant on the straight edge response for small central obscurations.

Figure 3.26 shows the results for various values of γ, the ratio of the inner to the outer radii of the annuli [3.10]. Unfortunately, we can see that any obscuration of the aperture degrades the performance. However the increase in depth of focus may warrant the use of a small obstruction. As the side lobes with circular lenses are so small it still seems likely that there is some apodisation which would result in an improved straight edge response.

3.6   The image of a phase edge

We have just discussed the image of a strong amplitude object and so we now move on to discuss the image of a strong phase object such as a phase edge where the phase change is abrupt and not small. The image of such an object is of considerable importance in cell sizing and counting in biology and linewidth measurement in integrated circuit technology, where the edge of the cell or the metallisation may be thought of as the phase step.

We consider an object whose amplitude transmittance is alternatively $\exp j\phi_1$ and $\exp j\phi_2$, which may be written

$$t(x) = \exp j\phi_1 \left[\frac{(1 + \exp j\Delta\phi)}{2} - \frac{2}{\pi}(1 - \exp j\phi) \sum_{n=1}^{\infty} \frac{(-)^n}{2n-1} \cos(2n-1)\theta \right]$$

(3.63)

where $\Delta\phi = \phi_2 - \phi_1$ and θ is as defined in equation (3.57).

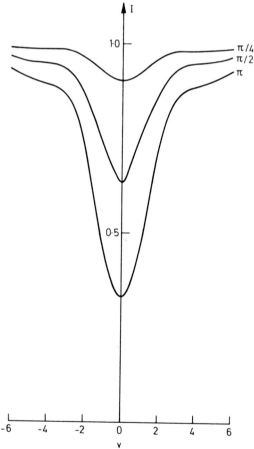

FIG. 3.27. The intensity in the image of a phase edge in a conventional scanning microscope.

Following the methods of the previous section we are able to write for the partially coherent conventional microscope [3.12]

$$I(x) = \left(\frac{1 + \cos \Delta\phi}{2}\right)C(0;0) - \frac{4 \sin \Delta\phi}{\pi} \, \mathrm{Im}\,\{S_1\}$$

$$- \frac{4}{\pi^2}(1 - \cos \Delta\phi)[S_2 + S_3 + 2(S_4 + S_5)] \qquad (3.64)$$

where the Ss are given by equation (3.61).

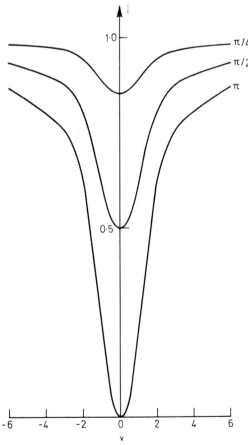

FIG. 3.28. The intensity in the image of a phase edge in a confocal scanning microscope.

The expressions are again much simpler for the confocal microscope and may be written

$$I(x) = \left| \left(\frac{1 + \exp j\Delta\phi}{2} \right) C(0) - \left[\frac{2(1 - \exp j\Delta\phi)}{\pi} S_1 \right] \right|^2. \qquad (3.65)$$

The responses are plotted in Fig. 3.27 and 3.28. We see that the phase edge manifests itself as a central dip in intensity which in all cases gives greater contrast for the confocal microscope. The extreme case of a 180° phase step gives a limiting central intensity of 0·333 in the conventional microscope compared with zero in the confocal instrument. The central intensity in the confocal case may be written as

$$I(0) = \cos^2 \left(\frac{\Delta\phi}{2} \right) \qquad (3.66)$$

which becomes smaller as $\Delta\phi$ increases as indicated in Fig. 3.28 reaching zero for $\Delta\phi = \pi$.

3.7   Depth discrimination in scanning microscopes

A large depth of field is often a desirable property in a microscope, such as when observing rough surfaces. Equally, when observing thick biological specimens in transmission it is often useful to limit the depth of field to avoid confusion in interpretation of the micrographs.

   If we observe the image of a single point in a conventional or Type 1 scanning microscope as the object is taken out of the focal plane we find that the image broadens and that the axial intensity decreases. Either of these quantities may be taken as an indicator of depth of field.

   The intensity variation near the focus of a lens may be written as

$$I(u, v) = |h_1(u, v)|^2 \qquad (3.67)$$

where h_1 is the impulse response of the lens and u, v the optical coordinates (equations 2.28 and 2.37), a constant multiplier being neglected. For the case when the lens has a circular pupil, Fig. 2.3 shows contours of equal intensity in the (u, v) plane. These results also show the intensity variation in the image plane (v is now proportional to the distance in the image plane) of a point object placed a normalised distance u from the focal plane of the lens. For a given object position the axial intensity in the image is given by

$$I(u, 0) = \left(\frac{\sin u/4}{u/4} \right)^2. \qquad (3.68)$$

In a scanning microscope the intensity in the image depends on the properties of both the objective lens and the collector lens. Assuming that the impulse response of the latter is h_2, the intensity is given by (equation 3.35)

$$I(u, v) = |h_1(u, v)h_2(u, v)|^2. \tag{3.69}$$

If both lenses are equal the intensity is simply the square of that in the Type 1 microscope. For two circular pupils the contours of Fig. 2.3 are still valid except that the value of the intensity of the contour is squared. The axial intensity varies as

$$I(u, 0) = \left(\frac{\sin u/4}{u/4}\right)^4, \tag{3.70}$$

which shows that if depth of field is defined in terms of the fall-off in maximum intensity for a point image, the depth of field for the confocal microscope is reduced relative to that of a conventional microscope, the difference not being very great.

There are, however, other associated properties which may be of practical importance. For instance we may investigate the variation in the integrated intensity for the image of our point source, which is a measure of the total power in the image [3.13]. This tells us how our microscope discriminates against parts of the object not in the focal plane. We define the integrated intensity as

$$I_{int}(u) = 2\pi \int_0^\infty I(u, v)v \, dv. \tag{3.71}$$

For the Type 1 microscope we know from Parseval's theorem that this is equal to the integral of the modulus squared of the effective (that is the defocused) pupil function, and since the effect of the defocus is merely to introduce a phase factor which disappears when the modulus is taken we come to the conclusion that the integrated intensity does not fall off in the Type 1 microscope. We may argue this equally from conservation of energy. Since our function $I(u, v)$ for the Type 1 microscope is the same as the intensity near the focus of a lens the integrated intensity is proportional to the power crossing any plane perpendicular to the optic axis, and this must be constant.

There is thus no discrimination of this kind in the Type 1 or conventional microscopes, but this is not as bad as it seems as the defocused image eventually becomes a constant background which is rejected by the observer, although of course it does reduce contrast.

Turning now to the confocal microscope we have, using the expression for the intensity in the focal region of a single lens (equation 2.40),

$$I_{int}(u) = 2\pi \int_{0}^{\infty} (C^2(u, v) + S^2(u, v))^2 v \, dv. \qquad (3.72)$$

This integral is plotted in Fig. 3.29 where we have normalised it to unity in the focal plane. The integrated intensity falls off monotonically, reaching the

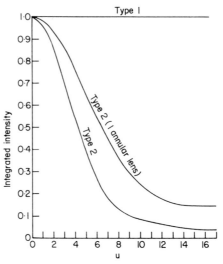

FIG. 3.29. The variation in the integrated intensity as a function of distance from the focal plane.

half power point at a distance from the focal plane of 0.70λ for a numerical aperture of unity. For large values of u the integrated intensity falls off according to an inverse square law, as may be shown by considering the geometrical optics approximation.

We have already discussed the use of annular lenses in scanning microscopes earlier in this chapter and it is therefore of interest to examine the effect of using one annular lens on the depth of field of the microscope. The variation in the intensity in the image of the single point is now exactly as for a conventional microscope (equation 3.5) as the impulse response of the annular lens does not vary along the axis. It should be noted that we have not said that the variation in the breadth of the point image is as in the conventional microscope: these are difficult to compare as the intensity variations differ in shape. The integrated intensity may again be computed, in this case from

$$I_{\text{int}}(u) = 2\pi \int_0^\infty (C^2(u, v) + S^2(u, v))J_0^2(v)v \, dv, \qquad (3.73)$$

and this is also shown in Fig. 3.29 again normalised to unity at the focal plane. The curve decreases monotomically, but the discrimination is not now so great as for the microscope with circular pupils. The integrated intensity has fallen to one half at a distance of $1 \cdot 05\lambda$ for a numerical aperture of unity.

We have seen that using one annular lens in a confocal microscope reduces the discrimination against objects outside the focal plane. If two annular lenses are used we would expect this discrimination to be further reduced. In the limiting case as the cross-section of the annulus is very small the integrated intensity becomes constant as the impulse response does not then vary along the axis. This is consistent with the claim of increased depth of field in a confocal microscope with two annular lenses [3.14].

3.8   Contrast mechanisms in confocal microscopy

In sections 3.1 to 3.6 we considered imaging of a thin object whose amplitude and phase effect on the transmitted (or reflected) beam is completely characterised by a complex function of position $t(x, y)$. Let us further consider two particular objects of this type. A linear variation in phase, as would be produced by a wedge of dielectric viewed in transmission or a sloping surface in reflection, has a single spatial frequency in its spectrum so that the image consists of a constant intensity the strength of which is given by $C(m, m)$. This imaging of phase gradients results from the fact that the refracted (or reflected) beam tends to miss the collector lens. The transfer function $C(m, m)$ falls off quicker with increasing slope in the confocal microscope than in a conventional microscope with equal pupils so that contrast resulting from variations in this phase gradient are imaged more strongly in the confocal system. The second example is a surface of small cosinusoidal height variations (equation 2.84). If b is the amplitude of the oscillations there is no image at the spatial frequency of the oscillations in an aberration free system. The image variations at twice this spatial frequency are of strength b^2 and are also proportional to the value of $C(m; -m)$ $- C(2m; 0)$. For an incoherent conventional microscope this vanishes as might be expected, but it also vanishes for conventional microscopes with equal pupils. For confocal microscopes it does not vanish and hence an image is formed. There are thus some differences in the contrast produced in confocal and conventional microscope images.

However the most important differences result from the depth discrimination properties described in section 3.7. If a reflection object has height variations of sufficient magnitude to result in a change in the spread function then the signal in a confocal microscope will vary accordingly. Imaging of this type is very difficult to analyse because the spread function is not spatially invariant. However if the height is only slowly varying diffraction by the object may be neglected and the image signal in a confocal microscope will result entirely from the depth discrimination. Consider an object consisting of a perfect reflector in a reflection microscope. The Fourier transform of the reflectance is simply $\delta(m)\,\delta(n)$ and consequently the signal is just (equation 3.24) $C(0;0)$. For a conventional microscope we can see from equation 3.26 that the signal is independent of focus position. But for a confocal microscope we have (equation 3.42)

$$I = \left| \int\int_{-\infty}^{+\infty} P_1(x, y) P_2(-x, -y)\,\mathrm{d}x\,\mathrm{d}y \right|^2 \tag{3.74}$$

or for two circular pupils

$$I(u) = \left| \int_0^1 \exp ju\rho^2\rho\,\mathrm{d}\rho \right|^2 \tag{3.75}$$

$$= \left(\frac{\sin u/2}{u/2} \right)^2. \tag{3.76}$$

So if the object is displaced from the focal plane in either direction the signal falls. It should be noticed that $C(0;0)$ is the intercept on the defocused transfer function (Fig. 3.12), but it should be remembered that this figure is for a one-dimensional model.

If the object is mounted with its normal slightly away from the optic axis, some parts will appear out of focus in a conventional or Type 1 image, but in a confocal microscope the image is modulated by equation (3.76) so that only the part of the object in focus is imaged [3.15] as shown in Figure 5.2.

At high numerical apertures only a very small height variation is necessary to produce a substantial depth discrimination effect so that this is an important source of image contrast. For higher spatial frequencies such that diffraction at the object may not be neglected there will be a combination of depth discrimination and diffraction imaging. In particular there will be an interaction caused by the complex defocused transfer function resulting in phase imaging of height variations.

Similar effects occur in transmission microscopy, a slab of dielectric of varying optical thickness in the object plane producing a variation in signal resulting from the change in effective separation of the lenses.

3.9 Scanning microscopes with partially coherent effective source and detector

We should finally briefly mention that all the microscope systems we have dicussed so far are merely special cases of a more generalised scanning microscope, that is, one with a finite source and a finite detector. This system has been analysed in great detail by Sheppard and Wilson [3.16] but we will do no more here than discuss a few special cases.

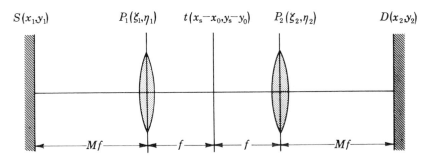

FIG. 3.30. The optical system of the generalised scanning microscope.

We consider the optical system of Fig. 3.30 and restrict ourselves to a one-dimensional analysis for simplicity, the two-dimensional result being an obvious extension. Using either the concept of mutual coherence [3.4] or the methods of Chapter 2 we may write the image intensity as

$$
I(x_s) = \int\limits_{-\infty}^{+\infty}\int\int\int S(x_1)h_1\left(x_0 + \frac{x_1}{M}\right)
$$

$$
\times h_1^*\left(x_0' + \frac{x_1}{M}\right)t(x_0 - x_s)t^*(x_0' - x_s)
$$

$$
\times h_2\left(x_0 + \frac{x_2}{M}\right)h_2^*\left(x_0' + \frac{x^2}{M}\right)D(x_2)\,\mathrm{d}x_1\,\mathrm{d}x_0\,\mathrm{d}x_0'\,\mathrm{d}x_2 \qquad (3.77)
$$

where S and D are the source and detector intensity sensitivities respectively.

We may now as a simple example consider the image of a point-like object. Then

$$I(x_s) = \left\{ \int_{-\infty}^{+\infty} S(x_1) \left| h_1\left(x_s + \frac{x_1}{M}\right) \right|^2 dx_1 \right\}$$

$$\times \left\{ \int_{-\infty}^{+\infty} D(x_2) \left| h_2\left(x_s + \frac{x_2}{M}\right) \right|^2 dx_2 \right\}$$

$$= \{ S(Mx_s) \otimes |h_1(x_s)|^2 \} \{ D(Mx_s) \otimes |h_2(x_s)|^2 \}. \qquad (3.79)$$

The image is clearly sharpest when both source and detector are points (the confocal case) and is degraded when either or both have finite size. As both become large the imaging becomes poor.

It is now possible to introduce the Fourier transform of t and obtain a general expression for the transfer function $C(m; p)$. However this is beyond the scope of our present intention; the conclusion of such an analysis being that in general the confocal arrangement possesses superior imaging properties.

3.10   The limitations of scalar diffraction theory

We have based our analyses so far on the application of the Kirchhoff diffraction formula in the paraxial limit. This approach necessarily involves certain simplifications and approximations apart from the use of the Kirchhoff boundary conditions. The paraxial condition essentially requires approximately $\tan x \sim x$ but this is clearly not appropriate if we wish to apply the results to microscopes involving high numerical aperture objectives. Further we have assumed light to be a *scalar* phenomenon, i.e. only the scalar amplitude of one transverse component of either the electric or magnetic field has been considered, it being assumed that any other components of interest can be treated independently in a similar fashion. This entirely neglects the fact that the various components are coupled to each other through Maxwell's equations and cannot strictly be treated independently.

To emphasise this we follow Hopkins [3.17] and consider a convergent spherical wavefront with the electric vector parallel to one principal section (Fig. 3.31). We would expect the electric vectors from A and B to combine vectorially at the focus to give a field less than their algebraic "scalar" sum. Conversely disturbances from the equivalent point of the meridian perpendicular to the plane of the diagram would be expected merely to add directly.

The rigorous solution of Maxwell's equations for such boundary value problems is rather complicated and has only been successful in a limited number of cases. A number of papers [3.18–3.21] have dealt with the behaviour in the focal region of a high numerical aperture system and do indeed predict considerable departure from the paraxial results.

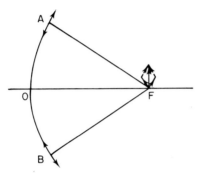

FIG. 3.31. The vector nature of the electric fields in the focal plane of a convergent spherical wavefront.

The important differences are as follows. The distribution of time-averaged electric energy density (which is the quantity which would be detected by, for example, a photographic emulsion) is not radially symmetric: the resolving power for measurements in the azimuth at right angles to the electric vector of the incident wave is increased relative to that of the paraxial theory, whereas in the plane containing the incident electric vector the resolving power is reduced. The minima of these distributions on one of the principal azimuths in the focal plane and along the optic axis are no longer zeros. The *total* time-averaged energy (electric plus magnetic) however *is* radially symmetric, but again the minima are not zeros. The electric field is not in general in the same direction as the incident electric field: it has both a cross and a longitudinal component. The magnitude of the time-averaged Poynting vector is also radially symmetric, the power flow forming closed eddies in the region of the focal plane.

These results are very important in certain applications, for example the analysis of telescope performance, although they are not, however, sufficient *per se* for microscope imaging, for in a microscope the object is illuminated by one system of high aperture *and observed* using another. It is quite surprising therefore that very little has been written on microscope imaging.

As an example Sheppard and Wilson [3.22] have calculated the image of one and two bright points in an opaque background for aberration-free microscope systems at high numerical apertures. They find that the image of a single point object remains radially symmetric even if the light source is plane polarised. For the conventional microscope the overall effects are quite small but the aperture of the condeser affects the image of the single point as well as determining the degree of coherence in the imaging process. The confocal microscope is found to give a slightly broader image than that predicted by paraxial theory and the depth of the minima are reduced. This is in agreement with the practical observations of Brakenhoff et al. [3.23].

The general conclusion of this discussion is that further work on the electromagnetic aspects of image formation would be immensely valuable but is likely to be very difficult. The fortunate fact is that for all its assumptions the scalar theory is able to predict to a surprising accuracy the effects that are observed in practice.

References

[3.1]   H. H. Hopkins and P. M. Barham (1950). *Proc. Phys. Soc.* **63**, 72.
[3.2]   T. S. McKechenie (1972). *Opt. Acta* **19**, 729.
[3.3]   H. H. Hopkins (1953). *Proc. R. Soc.* **A217**, 408.
[3.4]   H. H. Hopkins (1951). *Proc. R. Soc.* **A208**, 263.
[3.5]   M. Born and W. Wolf (1975). "Principles of Optics." Pergamon, Oxford.
[3.6]   R. A. Lemons (1975). Ph.D. Thesis, Stanford University.
[3.7]   H. H. Hopkins (1955). *Proc. R. Soc.* **A231**, 91.
[3.8]   C. J. R. Sheppard and T. Wilson (1980). *Phil. Trans. R. Soc.* **295**, 513.
[3.9]   C. J. R. Sheppard and T. Wilson (1979). *Appl. Opt.* **18**, 7.
[3.10] C. J. R. Sheppard and T. Wilson (1979). *Appl. Opt.* **18**, 3764.
[3.11] T. Wilson and J. N. Gannaway (1979), *Optik* **54**, 201.
[3.12] T. Wilson (1981). *Appl. Opt.* **20**, 3244.
[3.13] C. J. R. Sheppard and T. Wilson (1978). *Opt. Lett.* **3**, 115.
[3.14] C. J. R. Sheppard (1977). *Optik* **48**, 320.
[3.15] D. K. Hamilton, T. Wilson and C. J. R. Sheppard (1981). *Opt. Lett.* **6**, 625.
[3.16] C. J. R. Sheppard and T. Wilson (1978). *Opt. Acta.* **25**, 315.
[3.17] H. H. Hopkins (1943). *Proc. Phys. Soc.* **55**, 116.
[3.18] B. Richards (1956). *In* "Symposium on Astronomical Optics and Related Subjects" (Ed. Z. Kopal), p. 352. North Holland, Amsterdam.
[3.19] B. Richards and E. Wolf (1956). *Proc. Phys. Soc.* **B69**, 854.
[3.20] A. Boivin and E. Wolf (1965). *Phys. Rev.* **B138**, 1561.
[3.21] A. Boivin, J. Dow and E. Wolf (1967). *J. Opt. Soc. Am.* **17**, 1171.
[3.22] C. J. R. Sheppard and T. Wilson (1982). *Proc. R. Soc.* **A379**, 145.
[3.23] G. J. Brackenhoff, P. Blom and P. Barends (1979). *J. Microsc.* **117**, 219.

Chapter 4

Imaging Modes of the Scanning Microscope

4.1 General imaging considerations

It will be useful before we move on to discuss practical imaging schemes to return to our general partially coherent microscope of Chapter 3 and ask what form the transfer function should take in order that the variations in image intensity represent the required variation of object property. We recall that for an object which varies only in the x-direction, the image intensity may be written as

$$I(x) = \int\int_{-\infty}^{+\infty} C(m; p) T(m) T^*(p) \exp 2\pi j (m - p) x \, dm \, dp \qquad (4.1)$$

from which we see that

$$C(m; p) = C^*(p; m) \qquad (4.2)$$

as the intensity must be a real quantity.

We now choose to ignore diffraction effects and consider an object whose amplitude transmittance may be written as $t(x) = a(x) \exp j\phi(x)$. If

$$C(m; p) = 1 \qquad (4.3)$$

then

$$I(x) = |t|^2 \qquad (4.4)$$

which is often referred to as a perfect amplitude image. On the other hand if

$$C(m; p) = mp \qquad (4.5)$$

we obtain

$$I(x) = \left| \frac{dt}{dx} \right|^2 \tag{4.6}$$

which might be called differential contrast. We are also often interested in forming an image which depends on the difference in phase or amplitude, and in these cases if we set

$$C(m; p) = m + p \tag{4.7}$$

then

$$I(x) = 2a^2(x)\frac{d\phi}{dx} \tag{4.8}$$

which would represent differential phase contrast, whereas when

$$C(m; p) = j(m - p) \tag{4.9}$$

the image becomes

$$I(x) = 2\frac{d}{dx}\{a^2(x)\} \tag{4.10}$$

or differential amplitude contrast. This description is more appropriate in the study of weak objects for which we may write

$$a(x) = 1 + a_1(x) \tag{4.11}$$

and on the assumption that $a_1(x)$ is small, equation (4.10) becomes

$$I(x) = 4\frac{da_1}{dx} \tag{4.12}$$

which is indeed the differential of the amplitude.

We are now in a position to use some of these idealised results to make some general remarks on the form and symmetry of the $C(m; p)$ function. For differential phase contrast, for instance, it must be an odd function, being real for differential phase contrast and imaginary for differential amplitude contrast. These conditions, together with (4.1), dictate that the transfer function should possess the symmetry shown in Fig. 4.1. These results were, of course, obtained from an idealised system with no diffraction. The main effect of diffraction is to modify the transfer function such that there is a definite spatial frequency cut-off, but not to alter the over-all symmetry. It must be mentioned, however, that although many systems may possess the required symmetry, they do not all image equally well. The actual form of the transfer function is still of great importance.

We now move on to include the effects of diffraction, and to this end consider in detail the image of an object of the form

$$t(x) = \exp b \cos 2\pi v x \qquad (4.13)$$

where b is, in general, complex. The case when b is small is representative of many biological specimens, and under these circumstances the image intensity may be written

$$I(x) = C(0;0) + 2 \operatorname{Re} \{bC(v;0)\} \cos 2\pi v x$$

$$+ \tfrac{1}{2} bb^* C(v;v) + \tfrac{1}{2} \operatorname{Re} \{bb^* C(v; -v) \exp (4\pi j v x)\} \qquad (4.14)$$

where we have assumed the pupil functions to be symmetric.

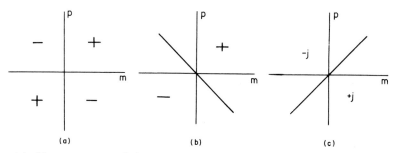

FIG. 4.1. The symmetry of the transfer function for (a) differential contrast, (b) differential phase contrast and (c) differential amplitude contrast.

If $|b|$ is not small, the image intensity is related in a complicated fashion to the real and imaginary parts of b. If we now specialise to the confocal rather than the conventional microscope, we have

$$C(m;p) = c(m)c^*(p) \qquad (4.15)$$

and equation (4.14) simplifies to

$$I(x) = |c(0) + bc(v) \cos 2\pi v x|^2 \qquad (4.16)$$

which is the modulus square of the amplitude image of the object.

If the microscope is in focus and b is wholly real we may take the square root of the intensity. The image then becomes a linear combination of the spatial frequencies, and is therefore highly suited to quantitative image processing. We notice here again that although the conventional and confocal microscopes have $C(m;p)$ functions which display the same symmetries, the quality of the final image depends crucially on the precise form of the transfer function.

Returning to the general case if $|b|$ is small, equation (4.14) reduces to

$$I(x) = C(0;0) + 2\operatorname{Re}\{bC(v;0)\}\cos 2\pi vx \qquad (4.17)$$

and the imaging depends only on the properties of the weak object transfer function $C(v;0)$. If the system is in focus, then only the real part of b is imaged. If on the other hand it is defocused, we have, as in Chapter 3, a complex $C(v;0)$. Introducing

$$C(v;0) = C_r + jC_i$$

and

$$b = b_r + jb_i \qquad (4.18)$$

we have

$$I(x) = C(0;0) + 2(b_rC_r - b_iC_i)\cos 2\pi vx \qquad (4.19)$$

and phase information is present in the image. Even if the phase variation in the object is small, a purely imaginary weak contrast transfer function $C(v;0)$ results in an image of the phase information only.

If we impose the condition that b must be small and expand equation (4.1), we can show that for a differential image the condition [4.1]

$$C(v;0) = -C(-v;0) \qquad (4.20)$$

must be satisfied, whereas for a standard non-differential image

$$C(v;0) = C(-v;0) \qquad (4.21)$$

as in the derivation of equation (4.14). In addition we have conditions for pure amplitude contrast

$$C(v;0) = C^*(-v;0) \qquad (4.22)$$

and for pure phase contrast

$$C(v;0) = -C^*(-v;0). \qquad (4.23)$$

We notice that these conditions obey the symmetry requirements of Fig. 4.1. A further requirement follows from equations (4.2) and (4.20)–(4.23), that $C(0;0)$ must be zero for either pure differential contrast or pure phase contrast.

4.2   The dark-field and Zernike microscope arrangements

4.2.1   *The dark-field microscope*

This is a simple method of examining low contrast specimens, and consists of obscuring the passage of light directly from the source to the image plane.

Thus the only light which reaches the image is that which has been diffracted or scattered by detail in the object, and it is only this detail which is observed.

The conventional dark-field microscope employs an annular condenser and a full objective. A similar arrangement may also be used in the scanning microscope. For the sake of simplicity, we idealise the pupil functions such that the condenser (collector) pupil is unity over a narrow annular region and zero elsewhere, and the objective pupil function is unity over a circular region of radius equal to the inner radius of the annular condenser pupil, such that when the two pupils are placed on top of each other there is no common area. This ensures that $C(0; 0)$ and $C(m; 0)$ are zero for both Type 1 and confocal microscopes.

The $C(m; p)$ surfaces, which were obtained from a geometrical interpretation of equations (3.28) and (3.42) are shown in Fig. 4.2(a) and (b). We can see that for the conventional dark-field microscope $C(v; -v)$ is zero, and so the image of a weak object (equations 4.13 and 4.14) is spatially constant. Thus single spatial frequency objects are not imaged. This has been noted previously by Burge and Dainty [4.2] who consider similar objects, but consisting of two spatial frequencies, and deduce that the image contains difference but not sum frequencies in addition to constant terms.

The confocal dark-field microscope behaves very differently as $C(v; v)$ is non-zero [4.1]. Thus single spatial frequencies are imaged [4.3], and the image intensity can now be written as

$$I(x) = |bc(v) \cos 2\pi v x|^2. \tag{4.24}$$

Thus we see that the dark-field method renders detail visible in terms of its diffraction pattern, and so is useful in examining weak objects. We can also see from equation (4.24) and the dark-field form of equation (4.14) that phase detail is also present in the image. This, then, is also a slightly more satisfactory way of imaging weak phase objects than the crude defocus technique of the previous chapter. It may be regarded as a limiting case of the more sophisticated Zernike phase contrast system, which modifies the bright-field image such that phase detail is imaged, rather than relying entirely on a dark-field image.

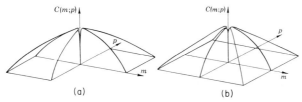

FIG. 4.2. The transfer $C(m; p)$ for dark-field imaging in the (a) conventional and (b) confocal microscopes.

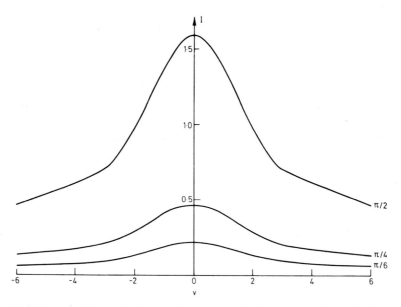

FIG. 4.3. The intensity in the image of a phase edge in a conventional dark-field microscope.

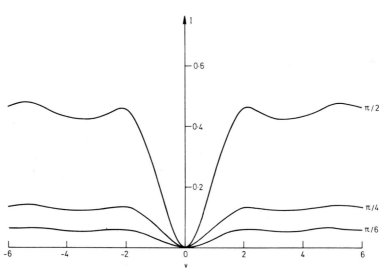

FIG. 4.4. The intensity in the image of a phase edge on a confocal dark-field microscope.

We have thus far been concerned with the images of weak phase objects. We conclude this section by discussing a strong phase object, the phase step. The images may be calculated by the methods of section 3.6 [4.4]. Figure 4.3 indicates that a phase edge is imaged in a conventional microscope by a local rise in intensity at the edge, the rise being greater the larger the phase step. We also note that far from the edge the intensity has fallen to a value of $\frac{1}{2}(1 - \cos \Delta\phi)$, which is to be expected, as our choice of an infinite annulus means that we are averaging over the entire field of view.

The confocal image is shown in Fig. 4.4. The most striking feature is that the intensity falls to zero at the edge regardless of the value of the phase step. This is a consequence of the coherent imaging of the confocal microscope, in which we essentially form an image of the amplitude transmittance of the object minus its mean value, and this function is zero at the edge. This, of course, is not the case for the partially coherent microscope, and as a result the image is indicated by a local rise in intensity.

For the sake of completeness, we should also mention that a very simple dark-field confocal microscope may be constructed by placing the limiting pinhole over the first dark ring in the Airy disc in the detector plane [4.5]. However, although this arrangement produces dark-field conditions, the transfer function is not the same as that obtained by conventional dark-field microscopy. Hence the two schemes produce slightly different images of the same object.

4.2.2   The Zernike phase contrast method

This was the first practical method of converting the phase differences suffered by the light passing through a phase object into observable differences in amplitudes [4.6]. The method consists of using an annular condenser and an annular phase ring in the objective. It is possible to increase the sensitivity of the method slightly by using a slightly absorbing phase ring [4.7].

Let us assume that the phase ring has a transmittance $c \exp (j\pi/2)$. We idealise the pupil functions such that the collector pupil is unity over a narrow annular region and zero elsewhere, and the objective pupil is $c \exp (j\pi/2)$ over a narrow annular region of equal diameter to that of the condenser: unity within the circular region inside the annulus and zero outside. The transfer function $C(m; p)$ is given for the conventional and Type 1 scanning microscope in Chapter 3 as the weighted area in common between the condenser pupil, the objective pupil displaced a distance m in spatial frequency space and the complex conjugate of the objective pupil displaced a distance p. We can break up the transfer function into three parts, the first being the area in common between the condenser pupil and the circular parts of the objective pupils, the second between the condenser pupil and the annular parts of the objective pupils, and the third part between the condenser pupil and the circular part of one objective pupil and the annular part of the other. The first part is real and its form is shown in Fig. 4.5(a). For $m = 0$ or $p = 0$ the transfer function is zero, but for small values of m and p, if they are both of the same sign, it quickly rises to a value corresponding to half the area of the annulus. Thereafter, for small p, the transfer function falls off as $\cos^{-1}(m\lambda f/2a)$. If m and p are of opposite sign, the transfer function is zero. The second part of the transfer function is also real (Fig. 4.5(b)). The three annuli intersect only if two annuli are coincident, that is, along the $m = 0$, $p = 0$, or $m = p$ lines. The value at $m = 0$, $p = 0$ is c^2 times the area of the annulus, this being normalised to unity in the diagrams. The third part of the transfer function is imaginary. In Fig. 4.5(c) we have shown the contribution from the convolution of a circle and an annulus for $m = 0$ and $p = 0$. There is also a very small non-zero value of the transfer function for other values of m and p if the width of the annulus is non-zero, but we have neglected this for the sake of simplicity.

We consider imaging of a weak object of the form given by equation (4.13)

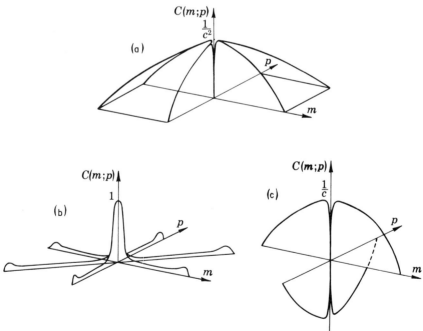

FIG. 4.5. The transfer function $C(m; p)$ for Zernike phase contrast imaging in a Type 1 scanning microscope.

such that equation (4.19) applies. The zero spatial frequency component in the image is taken to have a relative magnitude of unity, as given by Fig. 4.5(b). The real part of the cosinusoidal component is imaged with a strength given by the values of $C(m; 0)$ in Fig. 4.5(b). Similarly, the imaginary component is imaged with a strength given by the value of $C(m; 0)$ in Fig. 4.5(c). The properties of the image of a weak object are as follows:

i. The spatial frequency cut-off is the same as for a conventional microscope with full illumination, but the spatial frequency response for phase information is better in the Zernike arrangement than for a conventional amplitude contrast microscope.

ii. As Born and Wolf state, the intensity distribution produced in the image is directly proportional to the phase changes introduced by the object. This follows from equation (4.17), with the condition that the phase change is small.

iii. The contrast of the phase information is enhanced by a factor $1/c$ relative to the constant term.

iv. Weak amplitude information is imaged poorly, with fringes and low contrast.

The Zernike method suffers from a number of disadvantages, although it is a very convenient way of showing up phase information. The ideal geometry of the phase plate depends on the particular form of the specimen, there being four variables to consider: the diameter and thickness of the annulus, and the transmission and phase delay of the phase plate. De and Mondal [4.8] have discussed the effect of varying the thickness of the annulus with constant outer diameter. As might be expected, the cut-off frequency increases as the annulus width decreases. Mondal and Slansky [4.9] have considered what happens as the diameter of the annulus is varied with constant thickness. The cut-off frequency is greater for larger diameter annuli, but contrast is reduced.

For the confocal microscope, the transfer function is again made up of three parts [4.1]. If we represent the annular collector pupil by P_{2a} and the objective P_1 by $P_{1f} + jP_{1a}$, where P_{1f} represents the full part and P_{1a} the annular part, we can write

$$C(m; p) = \{P_{2a}(m) \otimes P_{1f}(m)\}\{P_{2a}(p) \otimes P_{1f}(p) + \{P_{2a}(m) \otimes P_{1a}(m)\}$$

$$\times \{P_{2a}(p) \otimes P_{1a}(p)\} + j\{P_{2a}(m) \otimes P_{1a}(m)\}\{P_{2a}(p) \otimes P_{1f}(p)\}$$

$$- \{P_{2a}(m) \otimes P_{1f}(m)\}\{P_{2a}(p) \otimes P_{1a}(p)\}. \tag{4.25}$$

The first two terms here representing the real part are shown in Fig. 4.6(a) and (b). We note that these pupil functions ensure that $C(0; 0)$, $C(m; 0)$, $C(p; 0)$, etc. are all zero in Fig. 4.6(b). The imaginary part is also made up of two components, each of opposite sign. If we assume that the annulus is narrow, the imaginary part is of greatest magnitude near to the m and p axes, and this is what we have shown, for simplicity, in Fig. 4.6(c).

The image is again composed of a constant background, a dark-field image and the phase contrast image. For a weak object, the image is precisely the same as a Type 1 or conventional Zernike arrangement. For a strong object, there are differences, but they become smaller as the thickness of the annulus becomes smaller, except for the superimposed dark-field image, which is always superior in the confocal microscope.

Again we conclude by discussing the image of a phase edge. The images are shown in Figs 4.7 and 4.8, and certain differences are noticeable between the

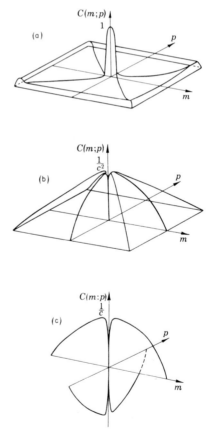

FIG. 4.6. The transfer function $C(m; p)$ for Zernike phase contrast imaging in a confocal scanning microscope.

conventional and the confocal images, in particular the high degree of fringing on the conventional image. This may be explained by considering the form of the image indicated by equation (3.64). The image we have here may be thought of as being a superposition of a phase given by the first two terms and a dark-field image given by the last term. Thus the extra fringing in the conventional image is a consequence of the form of the conventional dark-field image, which has a central maximum rather than the minimum of the confocal response. Nonetheless, the Zernike method is the first to give a steplike image. The intensity far from the edge is given by $(1 \pm \sin \Delta\phi)$, from which we can confirm that for low values of $\Delta\phi$, constant phase changes are not imaged well, and that such an edge would only be visible in terms of the

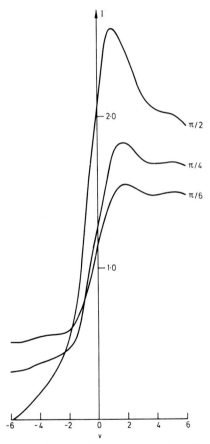

FIG. 4.7. The intensity in the image of a phase dege in a Type 1 Zernike phase contrast microscope.

fringing. We also note that, unlike the dark-field case, the method is sensitive to the direction of change of phase, i.e. the sign of $\Delta\phi$ is important. It is interesting that in the limiting case of $\Delta\phi = \pi$ (where the average value of the amplitude is zero), the image is exactly the same as would be obtained in a dark-field microscope.

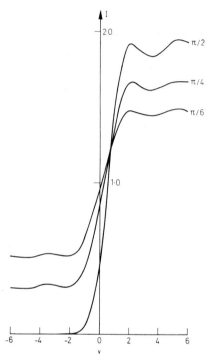

FIG. 4.8. The intensity in the image of a phase edge in a confocal Zernike phase contrast microscope.

4.2.3   The halo effect in dark-field and Zernike phase contrast microscopy

Our discussion so far has been concerned exclusively with the images of weak phase objects. While such objects are representative of a great many biological specimens, we must now turn our attention to strong objects, and the phase edge in particular. This is not a pure abstraction: it may, for example, reasonably model the edge of a biological cell.

The dark-field and Zernike schemes are two examples of a whole range of imaging techniques which rely on the presence of a spatial filter in the Fourier transform plane to remove or modify the zero spatial frequency component of the image [4.8]. All these methods have the disadvantage that the image of a sharp discontinuity exhibits a distinct halo, which may extend for many resolution elements [4.9]. The presence of this fringing is inherent in the method, as absolute phase is not imaged. Thus a phase edge, for example, would not be visible without fringing. We might think that the fringing is caused mainly by the annular region of the pupil, as might be expected for imaging with an annular lens, but in fact this is not the principal mechanism.

The haloes are produced irrespective of the degree of coherence of the imaging system. For the sake of clarity, then, we restrict ourselves to a coherent system, although this means that the results will generally not be applicable near the edge [4.10].

A straight-edge object is defined to have an amplitude transmittance

$$t(x', y') = 1; \quad x' > 0$$
$$= 0; \quad x' < 0; \quad \forall y' \tag{4.26}$$

for which

$$T(m, n) = \frac{1}{2}\left\{\delta(m) + \frac{1}{j\pi m}\right\} \tag{4.27}$$

and thus the image, which is only a function of x, becomes

$$I(x) = \frac{1}{4}\left| c(0) + \frac{1}{j\pi} \int \frac{C(m)}{m} \exp(2\pi jmx)\, dm \right|^2. \tag{4.28}$$

In a conventional coherent microscope, the coherent transfer function, $C(m)$, is given by the pupil function of the lens which for a Zernike system may be written as

$$C(m) = d \exp(j\pi/2) \quad 0 < |m| < a$$
$$= 1 \quad a < |m| < b \tag{4.29}$$

where b is related to the outer radius of the lens and a to the radius of the central phase disc, which has a transmittance d. By setting $d = 0$, we reduce to the case of the dark-field microscope [4.11]. We therefore have

$$I(x) = \frac{1}{4}\left| dj + \frac{2}{\pi}\{(dj - 1)\, \text{Si}\,(2\pi ax) + \text{Si}\,(2\pi bx)\} \right|^2 \tag{4.30}$$

and Si is the sine integral.

The image now consists of a fine structure of fringes superimposed on a slowly varying background, the scale of the fine structure being determined by the outer radius of the lens pupil. The limiting case as b becomes much larger than a describes the slowly varying background as

$$I(x) = \left\{1 - \frac{2}{\pi}\, \text{Si}\,(2\pi ax)\right\}^2; \quad x \neq 0. \tag{4.31}$$

The halo is therefore the result of diffraction by the finite size of the central disc. The extreme case of *a* tending to zero results in a halo that does not decay within the extent of the image, which then becomes a central dip on a uniform background. If we now move on to consider a phase edge which may be defined as

$$t(x', y') = \exp j\phi_1; \quad x' > 0$$
$$= \exp j\phi_2; \quad x' < 0 \tag{4.32}$$

we can combine the previous results to obtain an expression for the halo, in this case (when $b \gg a$) as

$$I(x) = \left| \cos\left(\frac{\Delta\phi}{2}\right) + \frac{2}{\pi d}\sin\left(\frac{\Delta\phi}{2}\right)\left[(dj - 1)\,\mathrm{Si}\,(2\pi ax) + \frac{\pi}{2}\right] \right|^2. \tag{4.33}$$

The essential features are sketched in Fig. 4.9, where the fringes have been omitted for clarity.

For the partially coherent Zernike arrangement employing an annular source and phase ring, the fringes are almost absent, and they could be further reduced by apodisation of the phase ring. The resolution is principally determined by the sum of the outer aperture of the annular source and the aperture of the objective. The width of the halo is determined by the minimum spatial frequency transmitted without change of phase, that is, by the difference between the outer aperture of the annular source and the inner aperture of the phase ring (or vice versa). In practice, we should aim for the width of the halo to be either as wide or as narrow as possible. If it is very wide, the image gives an intensity change proportional to the change of phase

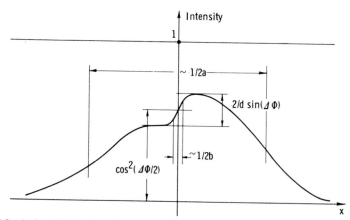

FIG. 4.9. A sketch showing the variation in intensity in the image of a phase edge in the Zernike phase contrast microscope.

superimposed onto the halo, whereas if it is very narrow, we obtain imaging somewhat similar to that in dark-field. If we choose to make the halo as wide as possible, this width is limited by three major factors:

i. The annular source must be of sufficient area to provide a suitable level of illumination. This is particularly important in conventional microscopes.

ii. The thickness of the annular phase ring must be sufficiently large compared to the annular source to allow easy alignment.

iii. There is a fundamental limit set by the requirement that the thickness of the annuli must be large compared to the wavelength.

In practice, the dimensions are usually such that the halo does not fill the entire field of view.

In phase contrast or dark-field, we are imaging a change relative to some average background amplitude. The effects which we have discussed are a result of the fact that the averaging is performed over some finite region of the object. The principles are the same for the case of edge enhancement using a split pupil in which one half experiences a phase change relative to the other, as the transition region must be of non-zero width and cannot be aligned exactly with the zero spatial frequency position. The Schlieren method, in which one half of the pupil is obscured, behaves similarly, as in practice it is impossible to obscure the zero spatial frequency component without also obscuring some of the lower spatial frequencies.

4.3 Interference microscopy

The basic criterion for a source in a scanning optical microscope is that it should be sufficiently bright to give a reasonable signal to produce a good picture. Hence a laser is now usually used. A further advantage of using a laser is that the coherence length of the beam makes the construction of an interference microscope relatively easy, as the optical path lengths in the two arms do not have to be made equal within such close tolerances.

We will begin by discussing the general scheme of Fig. 4.10, in which light transmitted through two dissimilar parallel optical paths illuminates two photodiodes which are arranged to give signals corresponding to the sum and difference of the amplitudes of the two beams [4.1]. The arrangement is analogous to the Mach–Zehnder interferometer. Thus, each detector essentially gives

$$I_\pm = |O \pm R|^2 \tag{4.34}$$

where O and R refer to the object and reference beam respectively. If we now electronically subtract these two signals, we are left with an interference term of the form

$$I_+ - I_- \sim \mathrm{Re}\,(OR^*) \tag{4.35}$$

Thus by careful choice of reference beam, we can image either the real or the imaginary part of the object amplitude transmittance.

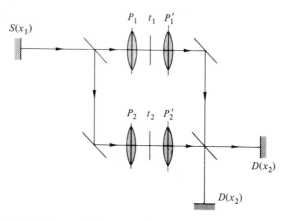

FIG. 4.10. The optical system of an interference microscope.

In order to discuss this scheme in more detail, we include a transversely incoherent source of finite distribution $S(x_1)$ and an incoherent detector of sensitivity $D(x_2)$. The lenses in the first path have point spread functions h_1 and h_1'; similarly, h_2 and h_2' in the second path. Using the methods of Chapter 3, and restricting ourselves to the one-dimensional case, we obtain for the image intensity

$$I_\pm(x) = \sum_{i=1}^{2} \sum_{j=1}^{2} \int\int\int\int_{-\infty}^{+\infty} (\pm)^{i-j} h_i\left(\frac{x_0 + x_1/M}{\lambda d}\right) h_j^*\left(\frac{x_0' + x_1/M}{\lambda d}\right)$$

$$\times\, t_i(x - x_0)t_j^*(x - x_0')h_i'\left(\frac{x_0 + x_2/M}{\lambda d}\right) h_j'^*\left(\frac{x_0' + x_2/M}{\lambda d}\right)$$

$$\times\, S(x_1)D(x_2)\,\mathrm{d}x_1\,\mathrm{d}x_0\,\mathrm{d}x_0'\,\mathrm{d}x_2 \tag{4.36}$$

where t_1 and t_2 are the amplitude transmittances of the two objects. However, we are particularly interested in the case when t_2 is a constant complex quantity w, as by carefully choosing this dummy object we can select the part of the object transmittance that is imaged. If we subtract the sum and difference signals, the resulting image in the Type 1 scanning microscope is

$$I(x) = 4 \operatorname{Re}\left[w^* \int_{-\infty}^{\infty} \left\{ h_2^*\left(\frac{x_0'}{\lambda d}\right) h_2'^*\left(\frac{x_0' + x_2/M}{\lambda d}\right) dx_0' \right\} \right.$$

$$\left. \times \left\{ \int_{-\infty}^{\infty} h_1\left(\frac{x_0}{\lambda d}\right) t_1(x - x_0) h_1'\left(\frac{x_0 + x_2/M}{\lambda d}\right) dx_0 \right\} dx_2 \right] \quad (4.37)$$

whereas for a confocal scanning microscope

$$I(x) = 4 \operatorname{Re}\left[\left\{ w^* \int_{-\infty}^{\infty} h_2^*\left(\frac{x_0'}{\lambda d}\right) h_2'^*\left(\frac{x_0'}{\lambda d}\right) dx_0' \right\} \right.$$

$$\left. \times \left\{ \int_{-\infty}^{\infty} h_1\left(\frac{x_0}{\lambda d}\right) h_1'\left(\frac{x_0}{\lambda d}\right) t_1(x - x_0) dx_0 \right\} \right]. \quad (4.38)$$

The expression for the Type 1 microscope is quite complicated. For the confocal microscope, however, the first integral is a constant giving the modulus and phase of the reference beam, while the second is the convolution of $h_1 h_1'$ with the object transmittance. If the lenses in the first optical path are aberration-free such that h_1 and h_1' are real, we can image the real or imaginary part of the object transmittance by choosing w such that the expression in square brackets is either real or imaginary. This is true even with strong contrast objects, as is the fact that the image is a linear function of the amplitude transmittance of the object. It is of great practical importance that the image depends only upon the amplitude of the reference beam at the detector pinhole. Thus, the shape of the reference beam is immaterial.

In practice, we might arrange for the elements in the two paths to be identical, in which case the equations become much simpler. If we now introduce the Fourier transforms of the object transmittances, we can write the image in the form

$$I_{\pm}(x) = \int_{-\infty}^{+\infty} C(m; p)\{T_1(m) \pm T_2(m)\}\{T_1^*(p) \pm T_2^*(p)\}$$

$$\times \exp 2\pi j(m - p)x \, dm \, dp \quad (4.39)$$

where $C(m; p)$ is the transfer function for a single optical path. If the second object has constant transmittance w, we obtain

$$I_\pm(x) = \left| \int\int_{-\infty}^{+\infty} C(m; p)T_1(m)T_1^*(p) \exp 2\pi j(m - p)x \, dm \, dp + |w|^2 \right.$$

$$\pm 2 \operatorname{Re} w^* \int_{-\infty}^{\infty} C(m; 0)T_1(m) \exp (2\pi jmx) \, dm.$$

We note that the important interference term is modified by the transfer function $C(m; 0)$. This is the case for any source and detector distribution. Reference [4.1] shows that this reduces to the corresponding transfer function for the non-interference case. However, as we know that $C(m; 0)$ is identical for Type 1 and 2 microscopes with circular aberration-free pupils, so these systems should behave identically. In the Type 1 case it is important that the lense in the two paths are identical and that the two beams are accurately aligned over the detector surface. With the Type 2 microscope, on

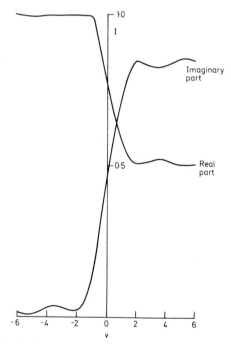

FIG. 4.11. The intensity in the image of a phase step from 0 to $\pi/3$ in an interference microscope.

the other hand, the amplitude and phase of the reference beam at the point detector are the only important properties, and the alignment of the system is much less critical.

We recall from Chapter 3 that the use of one full lens and one annular lens results in the enhancement of the $C(m; 0)$ function at high spatial frequencies. Such a combination has obvious advantages. If we choose to employ such a system, the relative size of the annulus is important. If the annular lens is larger than the full lens, such as in dark-field, and if this combination is used in the reference path, there will be, for constant t_2, no reference beam. On the other hand, if it is used only in the object arm, we can build a dark-field interference microscope which, unlike conventional dark-field instruments, can image single spatial frequencies.

If we now return to our phase step object, we see that we may produce an image which depends on the absolute value of ϕ_1 and ϕ_2 rather than just their difference. In particular the real part of the phase step will be imaged as a step of height $\cos \phi_1 - \cos \phi_2$ and the imaginary part as a step of height $\sin \phi_1 - \sin \phi_2$. Figure 4.11 shows a typical image when $\phi_1 = 0$ and $\phi_2 = \pi/3$ for a microscope with two equal circular pupils. The response may also be improved slightly by introducing an annular lens into both paths, as this serves to enhance the higher spatial frequencies.

It is possible to extend the schemes we have discussed above to include two reference beams and two dummy objects. By carefully selecting the dummy objects, we can arrange to image the real and imaginary parts of the object simultaneously. These electrical signals could then be processed, and we could, for example, display an image in which intensity changes in direct proportion to phase changes in the object.

A practical arrangement for a reflection confocal interference microscope is shown in Fig. 4.12, in which the radiation is focused into a plane parallel beam before passing through the first beam splitter [4.12]. It is apparent

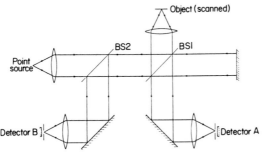

FIG. 4.12. A schematic diagram of a reflection-mode interference-mode confocal scanning microscope.

from equations (4.34) and (4.35) that the sum signal I_S consists of the normal confocal image $|O|^2$ superimposed on a constant $|R|^2$, while the difference signal, I_D leaves a pure interference image. Because this signal can be bidirectional, an electronic offset is added to it, producing, for display purposes, a signal which is always positive-going.

Figure 4.13 shows the image of an area of a TEM grid formed just from the signal I_A, showing a series of fringes representing variations in surface height superimposed on the confocal image. Local deformations of the fringes correspond to surface detail. Figure 4.14 shows a comparison confocal image of the same region. A deep scratch is clearly visible in both images.

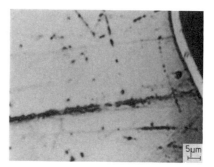

FIG. 4.13. The single detector inter-ference image, I_A, of part of a TEM grid using a He–Ne (wavelength 632·8 nm) laser and a 0·5 numerical aperture objective.

FIG. 4.14. Confocal image of the same region as Fig. 4.13.

The images which have been shown so far have been produced using a single detector. The effects of using the two-detector system will now be described. The interference image, I_B, from the second detector is similar to I_A, except that the fringes are displaced by half a period, so that the signal I_S produced by addition should exhibit no fringes. In practice small optimisa-tions of the relative gains and pinhole alignments are needed to achieve the optimum results shown in Fig. 4.15 which is, as predicted, similar to the confocal images of Fig. 4.14. Figure 4.16 then records the image from the difference signal I_D, with no further adjustments made to the system, and shows the fringe pattern without the superimposed confocal image. Even though as equation (4.35) predicts, the brightness of the fringes is modulated by the reflectivity of the object, their positions are now much more easily observed, allowing the surface topography to be deduced more precisely.

The arrangements discussed so far use a reference beam external to the object, but it is also possible to use a reference beam which passes through the specimen as well [4.1]. For example, in a differential interference microscope the same optics are used to focus light simultaneously on two adjacent points, the beam being split into two and recombined using birefringent elements. Let us assume that the objects as seen by the displaced beams are

$$t_1(x) = t(x - \Delta); \qquad t_2(x) = t(x + \Delta) \qquad (4.40)$$

which, by applying the shift theorem give for the Fourier transforms gives

$$T_1(m) = T(m) \exp(-2\pi jm\Delta); \qquad T_2(m) = T(m) \exp(2\pi jm\Delta). \quad (4.41)$$

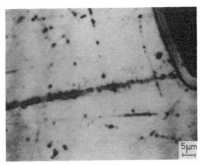

FIG. 4.15. The sum image, I_S, from the two-detector interference microscope. Note the similarity to Fig. 4.14 and the absence of fringes.

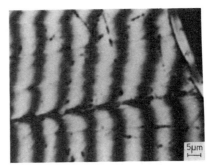

FIG. 4.16. Difference image, I_D, from the two-detector interference microscope, showing the fringe pattern without the superimposed confocal image.

Substituting in equation (4.39), and assuming that the signals are added with a phase difference 2ϕ, we have

$$I(x) = \int\limits_{-\infty}^{+\infty}\!\!\int C(m; p)T(m)T^*(p) \cos(2\pi m\Delta - \phi)$$

$$\times \cos(2\pi p - \phi) \exp 2\pi j(m - p)x \, \mathrm{d}m \, \mathrm{d}p \qquad (4.42)$$

where we have neglected a constant phase factor. If Δ is very small, the case when ϕ is zero reduces to the conventional image of t, whereas when $\phi = \pi/2$ we may write

$$I(x) = \int\limits_{-\infty}^{+\infty}\!\!\int C(m; p)T(m)T^*(p)4\pi^2\Delta^2mp \exp 2\pi j(m - p)x \, \mathrm{d}m \, \mathrm{d}p. \quad (4.43)$$

This may be written in the form of an effective transfer function

$$I(x) = \int\int_{-\infty}^{+\infty} C_{\text{eff}}(m; p)T(m)T^*(p) \exp 2\pi j(m - p)x \, dm \, dp \qquad (4.44)$$

where

$$C_{\text{eff}}(m; p) = 4\pi^2\Delta^2 mp C(m; p). \qquad (4.45)$$

This effective transfer function satisfies the conditions of equations (4.20), (4.23) and Fig. 4.1 for differential phase contrast imaging of weak objects. On the other hand, it also has the property that $C_{\text{eff}}(m; 0) = 0$, so that weak objects are not imaged at all. The imaging of weak objects may be improved by setting

$$\phi = \frac{\pi}{2} - \delta \qquad (4.46)$$

which modifies the effective transfer function to

$$C_{\text{eff}}(m; p) = C(m; p)(2\pi\Delta m + \delta)(2\pi\Delta p + \delta) \qquad (4.47)$$

The image of an object of the form

$$t(x) = 1 + jb_i \cos 2\pi vx \qquad (4.48)$$

is given by

$$I(x) = 1 - 2b_i\left(\frac{2\pi\Delta v}{\delta}\right)\sin 2\pi vx + \tfrac{1}{2}b_i^2 C(v; v)\left\{1 + \left(\frac{2\pi\Delta v}{\delta}\right)^2\right\}$$

$$+ \tfrac{1}{2}b_i^2 C(v; -v)\left\{1 - \left(\frac{2\pi\Delta v}{\delta}\right)^2\right\}\cos 4\pi vx. \qquad (4.49)$$

For imaging over a wide range of spatial frequencies we may take

$$\delta = 2\pi\Delta v. \qquad (4.50)$$

The weaker the object modulation, the greater is the range of spatial frequencies that is imaged linearly. It is therefore important to provide some means for the fine tuning of the relative phase of the two beams in the differential interference microscope.

As a second example of a system with a reference beam which passes through the object, we should mention the method of axial separation, in which the reference beam is focused at a point on the axis outside the specimen. Because this reference beam is defocused as it traverses the object, we may assume that the effective specimen seen by the reference beam is of unity magnitude. The image is then given by equation (4.39), and phase information may be obtained by suitable choice of the relative phase of the two signals.

4.4 Differential microscopy

Although a great deal of useful information may be obtained by studying a microscope image in which the contrast depends on the variation of either the amplitude (absorption) or phase of the amplitude transmittance of the object, it is often desirable to form an image which indicates how these quantities vary spatially across the object. This differential imaging is extremely useful, for example, in visualising the shape of an object or enhancing the contrast of point features.

There are many practical schemes of obtaining this contrast, and they may be divided into two main types. In the first, the differentiation is achieved by differentially detecting the radiation which has probed the object, whilst in the second a differential probe is employed.

4.4.1 Differentiation in the detector plane

We begin by describing a powerful scheme proposed initially by Dekkers and de Lang [4.13] to obtain phase information in the scanning transmission electron microscope without having to resort to the cruder methods involving the defocus and spherical aberration of the electron lenses. The basic principle may be explained by considering a weak object of the form of equation (4.13) in the STEM configuration of Fig. 4.17, such that the field just after the object may be written as

$$U(x, x_0) = (1 + b \cos 2\pi v(x_0 - x))h(x_0) \qquad (4.51)$$

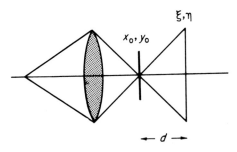

FIG. 4.17. The STEM configuration.

where x denotes the scan and h is the impulse response of the lens. The field at the detector is given by the Fourier transform of this expression, thus

$$U(x, \xi) = \int_{-\infty}^{\infty} U(x, x_0) \exp\left(\frac{-2\pi j}{\lambda d} x_0 \xi\right) dx_0 \qquad (4.52)$$

which for a symmetrical pupil gives

$$U(\xi, x) = P(\xi) + \frac{b}{2} P(\xi - \lambda dv) \exp(-2\pi jvx) + \frac{b}{2} P(\xi + \lambda dv) \exp(2\pi jvx)$$

$$(4.53)$$

which may be associated with three patches of light on the detector corresponding to an undeflected wave and two diffracted beams, one to the left and one to the right. This is depicted in Fig. 4.18, where the detector has also been included. As the object scans (x varies) the intensity in the detector plane will also vary owing to the interference between the beam deflected to the left and the undeflected beam and the beam deflected to the right and the undeflected beam. This is indicated by the shaded area in Fig. 4.18. If we assume an unaberrated pupil function, the intensity in the left area of interference is proportional to

$$1 + \text{Re}\{b \exp(-2\pi jvx)\} \qquad (4.54)$$

while that to the right is proportional to

$$1 + \text{Re}\{b \exp(2\pi jvx)\}. \qquad (4.55)$$

We can see from these two equations that by choosing b to be imaginary, the modulation in the two areas of interference are opposite in phase. This is to be expected, since a phase object does not alter the total amount of power transmitted. However, with an amplitude object the intensity modulations in the two areas are in phase with each other.

Dekkers and de Lang took advantage of this different behaviour of phase

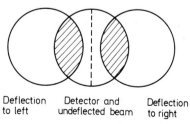

Deflection   Detector and   Deflection
to left   undeflected beam   to right

FIG. 4.18. The positions of the deflected beams on a split detector.

and amplitude objects to divide their detector into two semicircular areas as shown in Fig. 4.18. If the signals from the two halves are *added*, amplitude information is imaged, whereas if they are *subtracted*, phase information is present in the image. The form of equations (4.23), (4.54) and (4.55) suggest that phase structure appears differentiated in the final image whereas amplitude structure, of course, is not differentiated.

We may analyse this system further by considering the transfer function which may be written as

$$C(m; p) = \int\!\!\!\int_{-\infty}^{+\infty} D(\xi, \eta) P(\lambda dm - \xi, \eta) P^*(\lambda dp - \xi, \eta) \, d\xi \, d\eta \qquad (4.56)$$

where $D(\xi, \eta)$ represents the detector sensitivity.

The case when the two signals are added clearly results in the usual partially coherent imaging of the conventional microscope or the STEM. However, if we neglect aberrations, we see that when the signals are subtracted, $C(m; p)$ is a real, odd function with an hexagonal cut-off. The symmetry of Fig. 4.19(a) confirms that this arrangement gives differential phase contrast. The weak object transfer function $C(m; 0)$ may be evaluated in the case of a circular pupil together with a split detector of the same size as

$$C(m; 0) = \frac{2}{\pi} \{\cos^{-1} \tilde{m} - \tilde{m}(1 - \tilde{m}^2)^{1/2}$$

$$- \cos^{-1} 2\tilde{m} - 2\tilde{m}[1 - (2\tilde{m})^2]^{1/2}\}; \quad \tilde{m} < \tfrac{1}{2}$$

$$= \frac{2}{\pi} \{\cos^{-1} \tilde{m} - \tilde{m}(1 - \tilde{m}^2)^{1/2}\}; \quad \tilde{m} > \tfrac{1}{2} \qquad (4.57)$$

where $\tilde{m} = m\lambda d/2a$, and a is the radius of both the detector and lens pupil. This is plotted in Fig. 4.19(b). We might also consider the effects of using different detectors and in particular the wedge detector, defined by

$$D(\xi) = \text{const } \xi \text{ circ} \left(\frac{\xi}{a}\right). \qquad (4.58)$$

Differential phase contrast results again if the signal from the two halves of the detector are subtracted. Although a slightly superior image is obtained [4.14], the simplicity of the split detector together with the high-quality conventional image that is simultaneously available during the scanning process makes this a very versatile arrangement.

In practice, as the difference signal is bidirectional, a constant voltage is usually added to it such that zero difference signal gives approximately half full screen brightness. Figure 4.20(a) shows the image of an integrated circuit

formed by adding the signals from the detector halves. It is identical to a
conventional image, the bright regions corresponding to metallisation which
stands up above the silicon surface. Figure 4.20(b) shows the differential
phase image formed by subtracting one of the two detected signals from the
other. (Positive or negative contrast may be produced according to which
signal is the subtrahend.) This image shows a pronounced effect of relief and
considerable detail, particularly on the surface of the metallisation. The

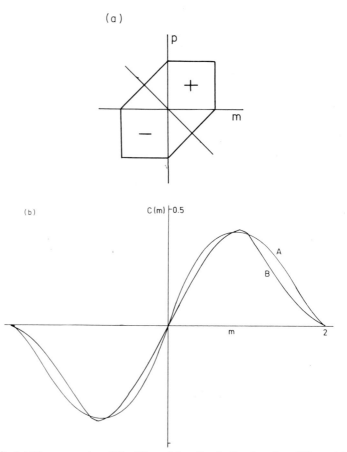

FIG. 4.19. (a) The symmetry of the $C(m; p)$ function indicating that differential phase
contrast imaging results. (b) The weak object transfer function for a split detector (A)
and a wedge detector (B).

FIG. 4.20. An integrated circuit viewed in reflection: (a) amplitude image, (b)
differential phase image and (c) a similar region viewed in a conventional Zeiss
microscope using Nomarski DIC.

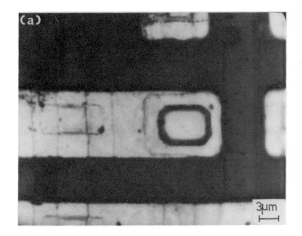

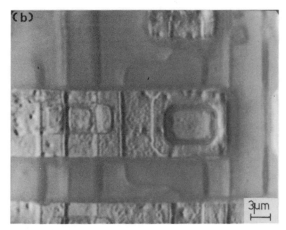

image from just one half of the detector consists of a conventional image with differential phase contrast superimposed, again leading to an impression of relief. Figure 4.20(c) represents a comparison image obtained with a conventional microscope using the Nomarski differential contrast technique described in the last section. The same optics are used to focus light simultaneously on two adjacent points of the object, the beam being split into two and recombined using birefringent elements.

A detailed comparison of these two techniques has been carried out [4.15] and the main differences are as follows. In the Nomarski technique, the image is a complicated mixture of different contrast mechanisms, the relative strengths of which can be altered by adjusting the compensator, while the split detector produces pure differential phase contrast. The Nomarski method also results in an asymmetrical response to phase gradients, unlike the split detector method. Further the Nomarski technique is unsuitable for examining birefringent objects.

It is, however, usual in scanning optical microscopy to include a lens after the object to collect the transmitted light onto a photodetector. We might ask what kind of image would result from the use of a coded detector in this case. We recall from the discussion in Chapter 3 that the two detector distributions which prove useful are the split detectors, by analogy with the conventional microscope. The two-point detector would consist of two pinholes placed symmetrically about the optic axis of the microscope. The transfer function for a microscope with a point source may be written as

$$C(m; p) = \int\limits_{-\infty}^{+\infty}\int F_D\left(\frac{\xi_1 - \xi_1'}{\lambda M d}\right)|P_2(\xi_1)|^2 P_1(\lambda dp - \xi_1')$$

$$\times P_1'(\lambda dm - \xi_1)\, d\xi_1\, d\xi_1' \qquad (4.59)$$

where F_D is the Fourier transform of the detector sensitivity function, which for a split detector, if we subtract the signals, is given by

$$F_D(v) \sim j\frac{1}{v} \qquad (4.60)$$

and again if we subtract the signals from a two-point detector

$$F_D(v) \sim j \sin \beta v \qquad (4.61)$$

where β is a constant proportional to the spacing between the point detectors. If β is small then

$$F_D(v) \sim jv. \qquad (4.62)$$

In both cases F_D is an imaginary odd function, and so, from equation (4.59) is $C(m; p)$. The symmetry is shown in Fig. 4.21, which indicates that this scheme results in differential amplitude contrast. We also note that the imaging is no longer coherent in the two point detector case.

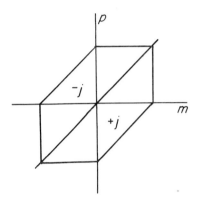

FIG. 4.21. The symmetry of the transfer function $C(m; p)$ indicating that differential amplitude contrast results.

4.4.2   Differentiation by a differential probe

In this approach, the optical arrangement in front of the object is modified in order that the object may be probed with what might be termed a differential spot. We have already discussed one such system in the section on interferometry, where birefringent elements were used to produce spatially separate spots to probe the object, thus producing differential phase contrast.

However, in this section we shall be more concerned with the effects of using lenses with coded pupil functions [4.14]. If the pupil function of the objective were an amplitude wedge, then we would probe the object with the differential of the Airy disc. In practice, it might be easier to use a split pupil, in which case the object is probed by

$$h = \frac{H_1(v)}{v} \tag{4.63}$$

in the differentiating direction, where H_1 is a first-order Struve function. Figure 4.22 illustrates these two forms of illumination.

We may now ask what kind of contrast would result from the use of such pupil functions. The easiest way to find out is to investigate the symmetry of the $C(m; p)$ function. We can see from equation (3.28) that for a conventional or Type 1 scanning microscope the transfer function is real and even irrespective of whether the condenser or objective is odd. Thus these microscopes do not give any differential contrast.

The confocal arrangement, however, behaves differently. Recalling that

$$c(m) = P_1(\lambda dm) \otimes P_2(\lambda dm) \qquad (4.64)$$

we find that $c(m)$ is a real odd function and that the cut-off in $(m; p)$ space is a square with the symmetry shown in Fig. 4.23 which results in what we have termed differential imaging. This is another instance of the power of coherent imaging in the confocal microscope: we have been able to incorporate the tools of coherent signal processing, but to a system having twice the spatial frequency bandwidth than is usual with such systems.

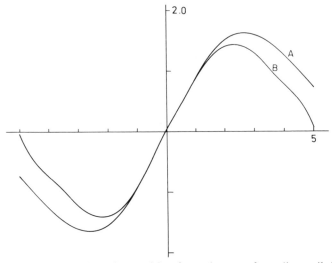

FIG. 4.22. The differential probe resulting from the use of a split pupil (A) and a wedge pupil (B).

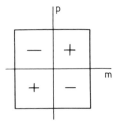

FIG. 4.23. The transfer function symmetry indicating that differential imaging results.

We have already indicated in Chapter 3 that the performance of a confocal microscope may be improved by employing lenses with annular apertures. Thus we now calculate the transfer function $c(m)$ for various combinations of pupils. A combination of one split detector and one annular lens gives

$$c(m) = \frac{2}{\pi} \{\cos^{-1} \tilde{m} - \cos^{-1} 2\tilde{m}\}; \quad \tilde{m} < \tfrac{1}{2}$$

$$= \frac{2}{\pi} \cos^{-1} \tilde{m}; \quad \tilde{m} > \tfrac{1}{2} \qquad (4.65)$$

One full lens and one split annular lens gives

$$c(m) = \frac{2}{\pi} \cos^{-1} \tilde{m} \operatorname{sgn}(\tilde{m}) \qquad (4.66)$$

while one full circular and one split circular lens has the transfer function of equation (4.57). These functions are plotted in Fig. 4.24. We may also calculate similar curves for the case of wedge pupils [4.14].

In confocal microscopy, the detected amplitude may be written

$$U = h_1 h_2 \otimes t \qquad (4.67)$$

and if one of the lenses has a wedge pupil function this becomes

$$U = h_1' h_2 \otimes t \qquad (4.68)$$

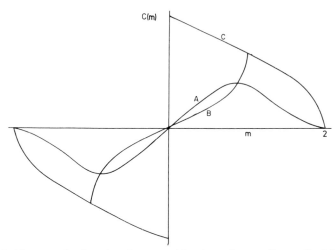

FIG. 4.24. The transfer function for a combination of one split pupil with one full circular pupil (A); one split pupil with one annular pupil (B); one full circular pupil with one annular split pupil (C).

the prime denoting differentiation. However, for the special case of $h_1 = h_2 = h$, this becomes

$$U = \tfrac{1}{2}h^2 \otimes t'. \tag{4.69}$$

That is, we form an image of the differential of the amplitude transmittance of the object.

We finally consider the image of a single point. Figure 4.25 shows the amplitude image for the various practically important cases, all of which have been normalised such that they have the same slope at the origin. We recall that in the non-differential coherent microscope a combination of one annular and one full lens resulted in good imaging. The same is true here, but it is now important that the annular lens should be responsible for the differentiation. We have also plotted the ideal differential response with one annular and one full lens in Fig. 4.25, that is

$$U = \frac{\partial}{\partial x}\left(J_0(x) \frac{2J_1(x)}{x} \right). \tag{4.70}$$

Upon comparison, we see that it is reasonably similar to the wedge annulus case.

Again, we may perform similar calculations for split pupils which display similar trends, the preferred combination being one full and one split annular lens.

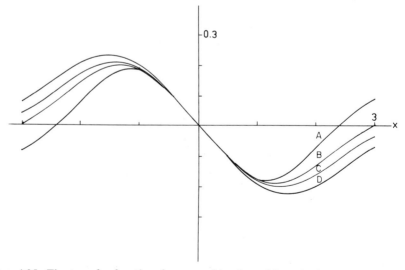

FIG. 4.25. The transfer function for a combination of lens 1 circular with lens 2 annular (A); the ideal differential image (B); lens 1 annular with lens 2 circular (C); two equal circular pupils (D).

4.5 Resonant scanning optical microscopy

If we place an object within an optical resonator, we can construct a multiple beam interference microscope which is very sensitive to changes in optical path [4.1]. A schematic diagram of the method is shown in Fig. 4.26. We calculate the effect on the signal of introducing an object into the resonator, and assume that the loss in the resonator is made up of a reflector power loss s_1 a diffraction loss s_2, and that the power transmission factors of the input and output mirrors are s_3 and s_4 respectively. Let us assume that the system is

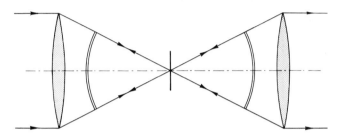

FIG. 4.26. Schematic diagram of the resonant scanning microscope.

at resonance with no object, so that the electric fields at the first mirror add in phase. If the incident field is E_o, the circulating field after reflection from the input mirror is E, and the signal field is E_{so}, we have, noting that the power loss per round trip is $2(s_1 + s_2) + s_3 + s_4$,

$$E_o s_3^{1/2} + E\{1 - (s_1 + s_2 + s_{3/2} + s_{4/2})\} = E \qquad (4.71)$$

and

$$E_{so} = E\left\{1 - \frac{s_1 + s_2}{2}\right\} s_4^{1/2}. \qquad (4.72)$$

Taking

$$s_3 = s_4 = s' \quad \text{and} \quad s_1 + s_2 = s \qquad (4.73)$$

we have for the signal intensity, if s is small,

$$I_{so} = \frac{I_o s'^2}{(s + s')^2}. \qquad (4.74)$$

Let us now assume that we have an object of transmittance t in the resonator. The fields still add in phase to give

$$E_o s_3^{1/2} + E\{1 - (s_1 + s_2 + s_{3/2} + s_{4/2})\}t^2 = E \qquad (4.75)$$

and

$$E_s = E\left(1 - \frac{s_1 + s_2}{2}\right) s_4^{1/2} t \tag{4.76}$$

whence, assuming $s' = s$ and that s is small,

$$I_s = \frac{4t^2 s^2 I_{so}}{\{1 - (1 - 2s)t^2\}^2}. \tag{4.77}$$

A typical figure of s obtainable with multilayer dielectric reflectors [4.16] of 2×10^{-3} gives a signal which varies as shown in Fig. 4.27. The behaviour is highly nonlinear, but for an almost transparent object

$$t = 1 - \delta \tag{4.78}$$

where δ is very small. We can write for the intensity

$$I_s = \left\{1 - 2\delta\left(1 + \frac{1}{s}\right)\right\} I_{so} \tag{4.79}$$

and so obtain a contrast enhancement of about 500 times. For a pure phase object

$$t = \exp j\theta \tag{4.80}$$

we obtain from equation (4.77)

$$I_s = \frac{I_{so}}{1 + (1 - 2s)s^{-2} \sin^2 \theta}. \tag{4.81}$$

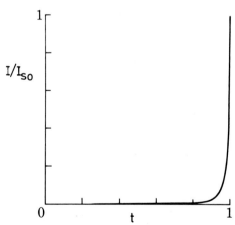

FIG. 4.27. Signal intensity in the resonant microscope as a function of object transmittance.

This is the normal expression for the fringes in multiple beam interferometry [4.7]. Our typical value of s gives a fringe half-width that, with visible light, corresponds to a path length of the order of one nanometre.

So far we have not mentioned the spatial resolution that can be obtained with a resonant microscope. We can obtain some idea of this by calculating the field in the focal plane of a concentric resonator. Assuming that only the lowest-order mode is present, the higher-order modes being attenuated by higher diffraction losses, an approximate expression for the amplitude at the mirrors in Fig. 4.26 is

$$\psi = J_0 \left\{ \frac{v_{01} r}{a(1 + \beta(i - j)/M} \right\} \tag{4.82}$$

where v_{01} is the first zero of the zero-order Bessel function, a is the radius of the mirrors, β is given by

$$\beta = -\frac{\zeta(\tfrac{1}{2})}{\sqrt{\pi}} = 0\cdot824 \tag{4.83}$$

where ζ is Riemann's zeta function, and M is related to the Fresnel number N by

$$\frac{M^2}{8\pi} = N = \frac{a^2}{2\lambda d}. \tag{4.84}$$

In practice one would use a very high Fresnel number, for which equation (4.82) reduces to

$$\psi = J_0 \left(v_{01} \frac{r}{a} \right). \tag{4.85}$$

In the focal plane we have an amplitude distribution given by

$$\psi(v) = \int_0^a J_0 \left(v_{01} \frac{r}{a} \right) J_0(v) v \, dv \tag{4.86}$$

where v is the optical coordinate. This integral may be evaluated [4.17] to give

$$I = \frac{J_0^2(v)}{\{1 - (v/v_{01})^2\}^2}. \tag{4.87}$$

As $v \rightarrow v_{01}$ both the numerator and denominator vanish and we obtain the limit by differentiation of these quantities

$$I_{\mathrm{Lt}(v \to v_{01})} = v_{01}^2 \, \frac{J_1^2(v_{01})}{4}. \tag{4.88}$$

The focal plane intensity is shown in Fig. 4.28. It is somewhat wider than the corresponding Airy disc, but exhibits extremely weak side lobes.

This, then, gives us an estimate of the resolution obtainable. If the object structure is coarse compared with this distribution, the radiation in the resonator will not be disturbed. However, to actually calculate the image of a particular object, we must calculate the modes of the resonator with an object present. The case of a microscope in which the radiation traverses the object twice is discussed in Chapter 6.

The resonant microscope has potential applications in studying small phase variations, and also small changes in height of a reflecting specimen. It is the coherence of the laser beam which allows the reflecting elements to be separated by such large distances. This geometry allows a large numerical aperture to be employed as the beam strikes the mirrors of the resonator normal to the coating.

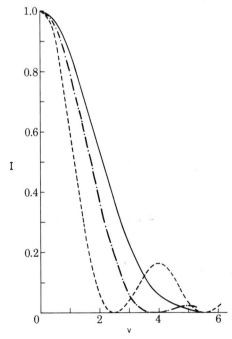

FIG. 4.28. Focal plane intensity distribution for a concentric resonator (lowest-order mode).

4.6   Synthetic aperture imaging

It has long been known that by laterally displacing one of the lenses in a microscope, the imaging properties are altered. This leads to easier visualisation of certain features of certain classes of objects. In order to understand this effect we consider weak objects and the changes in the $C(m; 0)$ function as the lens is moved. These functions are plotted in Fig. 4.29, and we see that in the final case, the transfer function passes spatial frequencies twice as high as in the in-line arrangement. This may be explained physically, as high spatial frequencies in the object result in large diffraction angles and as such will not be collected by the lens unless it is offset. The effect of this single sideband transfer function on the imaging is illustrated in Fig. 4.30, where we have redrawn the weak object transfer function as the sum of two functions. These may be recognised as resulting in differential phase contrast and dark-field images, but with the advantage of having twice the usual cut-off. Imaging of a similar form is obtained in the Schleiren system, where a knife edge is used to remove all the diffracted orders to one side of the zeroth, thereby resulting in a single sideband transfer function.

We might take advantage of the fact that the signal in a scanning microscope is obtained in an electrical form and can thus be processed before being displayed. If we consider the arrangement of Fig. 4.31 where we have one axial and two displaced lenses, we can combine the three signals to

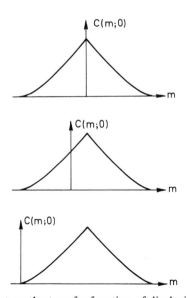

FIG. 4.29. The effect on the transfer function of displacing an objective lens.

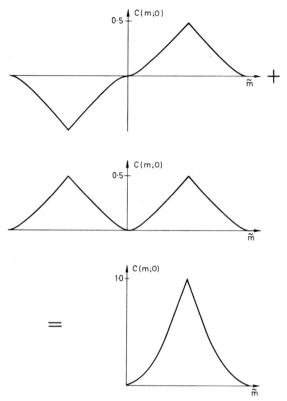

FIG. 4.30. A geometrical construction to illustrate that single sideband imaging results in differential phase contrast and dark-field imaging.

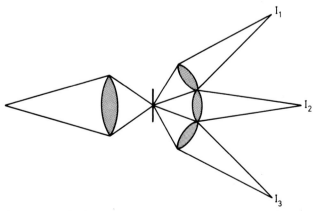

FIG. 4.31. Schematic diagram of a synthetic aperture imaging microscope.

produce various pure contrast forms. These are illustrated in Fig. 4.32. Adding the outer two gives pure dark-field imaging, while subtraction results in pure differential phase contrast. Summing all three signals clearly results in a conventional image. The great advantage of this approach is that in all cases the resolution is twice as high as usual but there is no need to use very expensive, high numerical aperture lenses. Although we have limited our discussion to three lenses it would be entirely possible to extend the argument to include the use of a multi-element fly's eye lens.

We may now go on to consider the images of strong objects. Fig. 4.33 indicates the form and symmetry of the $C(m; p)$ function when one of the outer signals is subtracted from the other, and confirms that in general differential phase contrast does result. However, recalling that $C(m; p)$ is the

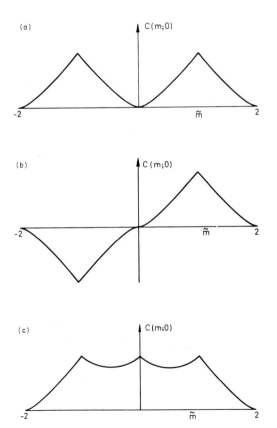

FIG. 4.32. The transfer functions resulting from various combinations of I_1, I_2 and I_3. (a) Darkfield; (b) differential phase contrast; (c) coventional imaging.

transfer function for the spatial frequency $(m - p)$ in the image, we can see that in the confocal case we have spatial frequencies twice as high as in the conventional and four times as high as in the non-offset coherent case. Thus the confocal arrangement is to be prefered.

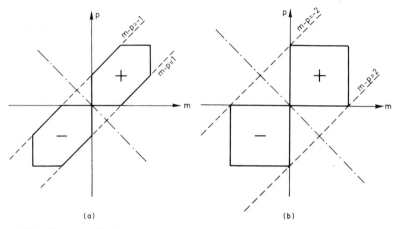

FIG. 4.33. The cut-off shapes in (m, p) space indicating the differences between (a) conventional and (b) confocal synthetic aperture imaging.

4.7   Stereoscopic microscopy

In a conventional stereoscopic microscope [4.18] the object is viewed from two slightly different angles. The stereoscopic effect comes from a combination of mechanisms, the most important of which is the parallax effect. This results in a lateral separation of the left- and right-hand images of each object point, the magnitude of which depends on the axial distance from the focal plane. As a secondary mechanism, an inclined surface appears with different intensity in the two images, resulting in highlighting. The shape of the object also differs in the two images according to the position in the object field and associated with this is a variation in the lateral parallax shift over the field.

Stereoscopic microscopes have a limited resolution, as it is not possible to arrange for two high numerical aperture optical systems to be placed side by side. The maximum numerical aperture which may be achieved in principle without immersion using a single objective with the pupil split into two semicircles, one being used for the left-hand image and the other for the right, is 0·707, rather than unity in a non-stereoscope microscope. In practice, however, it would be difficult to approach this limit closely. If a single objective with a numerical aperture of sin α is used in this way then the spatial frequency cut-off for each semicircular pupil is $2 \sin (\alpha/2)/\lambda$ rather than $2 \sin \alpha/\lambda$ for the whole objective.

In the confocal microscope, however, the spatial frequency cut-off for a reflection mode system with oblique illumination [4.19] is given by the maximum angle to the optic axis subtended at the object, rather than the semi-angle subtended by the objective at the object as in a conventional microscope. Thus if this instrument is used to examine an object, with semicircular halves of the pupil being used to produce left- and right-hand images (Fig. 4.34), the spatial frequency cut-off in the resulting image is maintained at $2 \sin \alpha/\lambda$. Points of the object far away from the focal plane

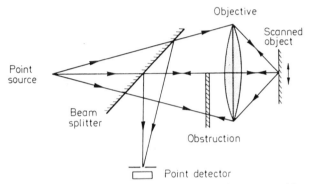

FIG. 4.34. Schematic diagram of a reflection confocal microscope with a half-plane obstruction.

undergo a lateral shift in the stereoscopic image pairs. The object may be both illuminated and detected using semicircular pupils, or alternatively, illuminated and detected using one semicircular and one circular pupil or vice versa, the latter method resulting in a weaker stereoscopic effect. If two semicircular pupils are employed, dark-field imaging results, so that in practice the pupils are obscured by a segment slightly smaller than a semicircle. A compromise must be made between high-quality imaging and a noticeable stereoscopic effect.

Figure 4.35 shows a high magnification stereoscopic image pair produced using a reflection microscope [4.20] and a semiconductor device as object. The objective had a numerical aperture of 0·5. The pair may be viewed by conventional methods, the separation being adjusted so that all object heights appear to be a positive distance in front of the observer. Figure 4.36 shows corresponding images produced with a large area detector. This arrangement is optically similar to a conventional stereoscopic microscope, except that in scanning methods there is no variation in object shape or in

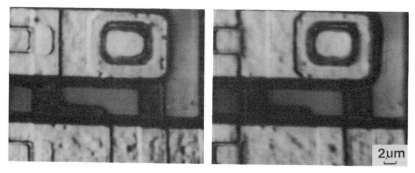

FIG. 4.35. High magnification reflection stereoscopic image pair formed using a confocal microscope. The area of each photograph is 30 μm × 23 μm.

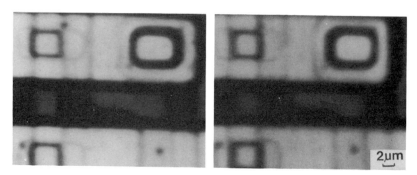

FIG. 4.36. A stereoscopic image pair formed without the detector pinhole, showing much reduced resolution.

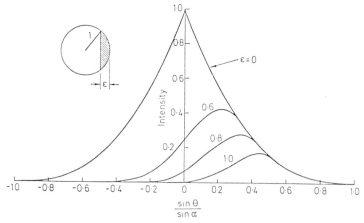

FIG. 4.37. The signal intensity for a perfect reflector at an angle θ for various obstructions ε.

parallax over the object field, as each point is brought to the centre of the field when imaged. In the resulting images, resolution is clearly much poorer than in those produced by the confocal system.

The confocal stereoscopic pairs exhibit lateral shifts corresponding to the axial distance from the focal plan. There is also a highlighting effect which results from a sensitivity to surface slope as shown in Fig. 4.37, where ε is the fraction of the pupil radius obscured, adding to the stereoscopic sensation. The intensity detected for a surface whose normal is at an angle θ to the optic axis is given by

$$I(\theta) = c\left(\frac{\sin\theta}{\sin\alpha}\right)c^*\left(\frac{\sin\theta}{\sin\alpha}\right) \qquad (4.90)$$

with

$$
\begin{aligned}
c(m) &= 0; & m &> 0 \\
&= \Lambda(m); & 1 &> m > \varepsilon/2 \\
&= \Lambda(m) - \Lambda(1 - \varepsilon + 2m); & 0 &< m < \varepsilon/2 \\
&= \Lambda(m) - \Lambda(1 - \varepsilon); & 0 &> m > -(1 - \varepsilon) \\
&= 0; & m &< -(1 - \varepsilon). \qquad (4.91)
\end{aligned}
$$

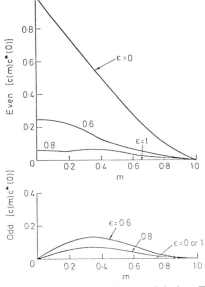

FIG. 4.38. The transfer function for weak object modulation. The even part results in amplitude imaging and the odd part in differential phase contrast.

The transfer function for weak object modulation is given by $c(m)c^*(0)$, which may be resolved into an even part, resulting in amplitude imaging, and an odd part giving rise to differential phase contrast. These are shown in Fig. 4.38. The differential phase contrast gives opposite contrast in the two images, thus producing a confusing stereoscopic image. The method as described so far is therefore only applicable for objects in which contrast comes from variations in amplitude or surface slope. For a weak phase object, on the other hand, a stereoscopic pair may be formed by inverting the contrast of one of the pair.

References

[4.1] C. J. R. Sheppard and T. Wilson (1980). *Phil. Trans. R. Soc.* **295**, 315.
[4.2] R. E. Burge and J. C. Dainty (1976). *Optik* **46**, 229.
[4.3] C. J. R. Sheppard and T. Wilson (1978). *Opt. Acta* **25**, 315.
[4.4] T. Wilson (1981). *Appl. Opt.* **20**, 3238.
[4.5] I. J. Cox, C. J. R. Sheppard and T. Wilson (1982). *Appl. Opt.* **21**, 778.
[4.6] F. Zernike (1935). *Z. Tech. Phys.* **16**, 454.
[4.7] M. Born and E. Wolf (1975). "Principles of Optics". Pergamon, Oxford.
[4.8] M. De and P. K. Mondal (1965). *J. Res. Nat. Bur. Stand.* **69C**, 199.
[4.9] P. K. Mondal and S. Slansky (1970). *Appl. Opt.* **9**, 1879.
[4.10] T. Wilson and C. J. R. Sheppard (1981). *Optik* **59**, 19.
[4.11] K. G. Birch (1968). *Opt. Acta* **15**, 113.
[4.12] D. K. Hamilton and C. J. R. Sheppard (1982). *Opt. Acta* **29**, 1573.
[4.13] N. H. Dekkers and H. de Lang (1974). *Optik* **41**, 452.
[4.14] T. Wilson and C. J. R. Sheppard (1980). Proc. 1980 Int. Opt. Computing Conf., SPIE, Vol. 232, p. 203.
[4.15] D. K. Hamilton and C. J. R. Sheppard (1984). *J. Microsc.* **133**, Pt 1, 27.
[4.16] C. J. R. Sheppard and R. Kompfner (1978). *Appl. Opt.* **17**, 2879.
[4.17] L. A. Vainstein (1969). "Open Resonators and Open Waveguide". Golan Press, Boulder, Colorado.
[4.18] D. Birchon (1961). "Optical Microscopy Technique". Newnes, London.
[4.19] J. P. J. Heemskerk, J. J. M. Braat and G. Bouwhuis (1981). *JOSA* **68**, 1407.
[4.20] C. J. R. Sheppard and D. K. Hamilton (1983). *Appl. Opt.* **22**, 886.

Chapter 5

Applications of Depth Discrimination

5.1 Introduction

We have already seen in Chapter 3 that an inevitable consequence of the point detector of the confocal microscope is that the microscope images detail only from the parts of the object around the focal plane. This depth discrimination property is not a feature of conventional or Type 1 scanning microscopes. The effect may be explained physically by considering the light from outside the focal plane of the collector lens, which forms a defocused spot at the detector plane (Fig. 5.1). The use of a central point detector results in a much weaker signal and so provides discrimination in the image against detail outside the focal plane.

The effect has been illustrated by deliberately mounting a planar microcircuit in a scanning optical microscope with its normal at an angle to the optic axis [5.1]. Figure 5.2(a) shows the Type 1 microscope image: only one portion, running diagonally, is in focus. Figure 5.2(b) is the corresponding confocal image: here the discrimination against detail outside the focal plane is apparent. The areas which were out of focus in Fig. 5.1(a) have now been rejected. Furthermore the confocal image appears to be in focus throughout the width of the visible band, demonstrating that the depth discrimination effect is dominant over the depth of focus. This is important, as it means that any detail that is imaged efficiently in a confocal microscope will be in focus.

5.2 Imaging with extended depth of field

For many purposes it is preferable to use a microscope with a small depth of field. One such application is the examination of a thick slice of biological tissue, where the section of interest may be surrounded by some transparent supporting medium. The depth discrimination property of the confocal scanning microscope makes it ideally suited for such applications. On the other hand a large depth of focus is often desirable, if, for example, we are interested in imaging a rough metallographic surface. An increased depth of field may be obtained in a conventional instrument, but at the price of poorer resolution. It is possible, however, to take advantage of the depth discrimination property and combine this with axial scanning of the object to produce a high resolution imaging instrument with vastly increased depth of field.

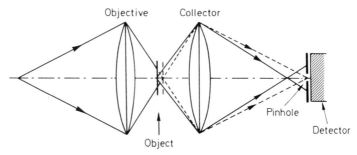

FIG. 5.1. Depth discrimination in the confocal microscope. When the object is in the focal plane (dashed lines) the transmitted light is focused on the pinhole. when the object is out of the focal plane, a defocused spot is formed at the pinhole, and the measured intensity is greatly reduced.

For example, if we had moved the specimen of Fig. 5.2 axially, we would have obtained a corresponding pair of images, but with a different portion of the object brought into focus [5.1]. Thus we can produce the image of a rough object such that all areas appear in focus by scanning the object in the axial (z-direction) with a amplitude sufficiently large for every part of its surface to pass through the focal plane. Needless to say, we must still scan in the x- and y-directions. The required frequency for the z-scan is determined from consideration of the horizonal (x-) and vertical (y-) resolution of the image: the z-scan frequency must be high enough to ensure that each picture point passes through focus at least once. If the optical resolution is such that there are N picture points per line of the raster and if the line scan time is T seconds, the axial scan frequency must be at least $(N/2T)$ Hz to meet this requirement. (Two picture points can pass through focus in each cycle of axial scan, one as the specimen travels each way.)

Figure 5.3(a), (b) shows the effect of mounting the microcircuit specimen on a piezoelectric biomorph and scanning it axially through a distance of about 20 μm [5.2]. It is clear that with axial scanning the Type 1 microscope gives the expected confused, blurred, out of focus image, whereas in the confocal mode only the in-focus detail has been accepted by the microscope, thereby producing an image which appears in focus across the whole field of view. This should be compared particularly with the conventional Type 1 image of Fig. 5.2(a). The main merit of this approach is that by introducing axial scanning we are able to produce "projected" images of rough specimens with resolution comparable with that produced in a conventional microscope by a high numerical aperture lens.

This technique is potentially extremely powerful, as it permits the depth of field in optical microscopy to be extended, in principle without limit, while still retaining high-resolution, diffraction-limited imaging. An experimental

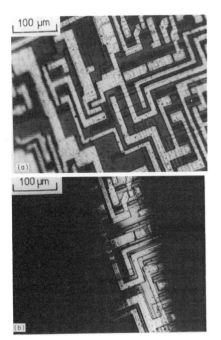

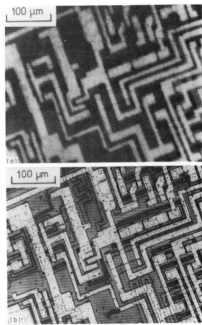

FIG. 5.2. (a) Conventional scanning microscope image of tilted specimen: the parts of the object out of the focal plane are blurred. (b) Confocal scanning microscope image of tilted specimen: only the part of the specimen in the focal plane is imaged strongly. 0·5 numerical aperture objective was used with a He–Ne (0·6328 μm wavelength) laser. A 10-μm diameter pinhole was used in the confocal mode.

FIG. 5.3. (a) Conventional scanning microscope image of tilted specimen with axial scan. (b) Confocal scanning microscope image of tilted specimen with axial scan.

extension of more than two orders of magnitude has already been achieved [5.3]. An example of this technique at high resolution is shown in Fig. 5.4 where the hairs on an ant's leg, with two hairs projecting to the left, have been imaged. The axial distance between the tips of these hairs is 30 μm, and Fig. 5.4(a) is an extended focus image which shows both excellently resolved

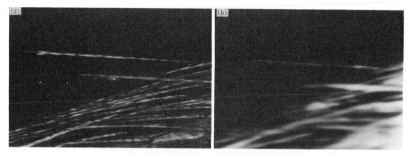

FIG. 5.4. The hairs on an ant's leg. (a) Extended focus image with 0·85 numerical aperture objective and He–Ne laser light. (b) Conventional scanning image with reduced (1/32) numerical aperture.

along their full length, as well as much detail on the leg itself. Figure 5.4(b), in which the microscope has been focused on the tip of the projecting hair, shows an attempt to increase the depth of focus in conventional microscopy by using a very low numerical aperture lens, but it is clear that as a result the resolution has suffered dramatically. Even so, nothing approaching the depth of field of the extended focus image has been achieved.

5.3   Surface profiling with the confocal scanning microscope

In the reflection confocal microscope, in the intensity $I(u)$, in the image of a point object placed on the optic axis a normalised distance u from the focal plane of the lens, is given by

$$I(u) = \left(\frac{\sin (u/2)}{u/2}\right)^2 \tag{5.1}$$

where u is measured in optical coordinates.

As the object passes through focus, the image intensity shows a sharp maximum. This result can be exploited to measure the surface profile of an object along a line scan (the x-direction, say). As the specimen is scanned in the axial (z-) direction, the position of the stage in each cycle of the scan where the intensity maximum occurs will depend on the height of the object's surface. For a depression in the surface, the maximum occurs when the stage is nearer to the lens, and for a projection when the stage is further away.

A measure of the stage position at the occurrence of the maximum in each cycle therefore gives a measure of the surface height of the specimen. In order to fully exploit the resolution of the microscope in the x-direction, the z-scan frequency must be such that at least one cycle of z-scan occurs for each resolution element in the x-direction: the object height is then being sampled once per picture point. In practice, however, to ensure accurate operation of the measuring system, the z-scan rate should be made many times this value.

It might be thought that a digital signal processing system would be required to analyse this data, but in fact a simple analogue arrangement is found to be perfectly satisfactory [5.4]. The aim of the system is to detect the time position of the maximum, and at this instant to sample the z-scan waveform and hold this value until the next cycle. This output will then give a measure of the surface profile.

Figure 5.5 shows a block diagram of the system, and Fig. 5.6 the waveforms at various points in it. The z-scan frequency will be assumed to be sufficiently high that over one cycle the object moves a negligibly small distance in the x-direction. Over one cycle, therefore, the detector output will vary as shown in Fig. 5.6(b), passing through its maximum twice as the object scans through focus in each direction.

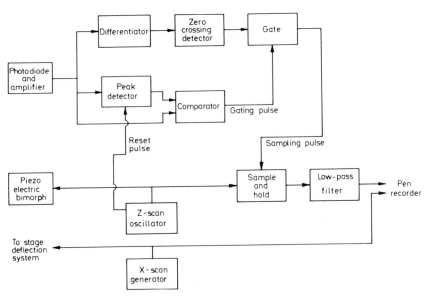

FIG. 5.5. Block diagram of the measurement system.

The detector signal is fed into a differentiator, whose output passes through zero at the signal maxima. However, there will be other turning points in the signal owing to the side lobes of the $I(u)$ variation, and it is necessary to ensure that all turning points apart from the large central maximum are ignored. To this end only the maximum in the second half cycle is used for the actual measurement, the amplitude V_p of the first maximum

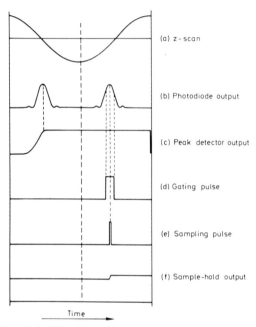

(a) z - scan

(b) Photodiode output

(c) Peak detector output

(d) Gating pulse

(e) Sampling pulse

(f) Sample-hold output

Time

FIG. 5.6. Waveforms in the measurement system.

being measured by a peak detector (which is reset to zero at the beginning of each cycle). The comparator then ensures that the trigger pulse to the sample-and-hold circuit is operative only when the signal level is greater than about $0.8 V_p$, that is, for a short period enclosing the second maximum. The object height is hence measured once per cycle of the z-scan. The sample-and-hold circuit is followed by a low pass filter which limits the bandwidth to that calculated from the x-scan rate, the x-scan amplitude, and the spot size— any higher frequency components must be due to noise from the sampling system. The filter output drives the vertical channel of a pen recorder, its horizontal channel being fed from the x-scan waveform.

The technique's potential may also be demonstrated with the microcircuit test object (Fig. 5.7). The profile of a metal strip is presented in Fig. 5.8. The information is presented as a three-dimensional map of the surface in isometric coordinates, with the x- and y-directions at 30° to the horizontal and the z-direction (specimen surface height) vertically on the plot. This method gives a fairly effective subjective impression of the surface shape [5.5]. The raster comprised 112 lines, which was optimum for clarity of display; and did not cause any serious loss of resolution as long as the y-scan amplitude was not greater than about 50 μm.

Figure 5.8 shows a profile across a metal conductor strip on the device. The shape of the strip is clearly defined and the difference between the surface texture of the metal and the surrounding semiconductor is apparent. The reflectivity of the metal and semiconductor is also quite different, but it does

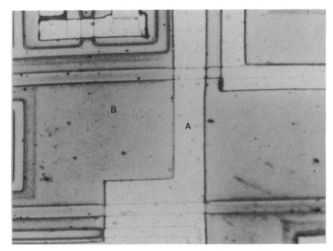

FIG. 5.7. Area of microcircuit. A, Metal; B, semiconductor.

FIG. 5.8. Profile of metal strip of microcircuit.

not affect the height measurement system. The fringing at the edge of the step is thought to be an artefact: theoretical calculations suggest that the higher numerical aperture of the imaging lens, the less this effect is noticeable. Nevertheless, the technique is a useful non-contacting method of surface profilometry. Depth variations of the order of $0.1\,\mu m$ have been clearly resolved [5.5].

5.4   The theory of image formation in extended field microscopy

In the previous chapters we have discussed the image formation properties of conventional and confocal microscopes where the object remains firmly in the focal plane. We have, however, seen in the previous sections of this chapter that high resolution images can be obtained by axial scanning in the confocal instrument. Now we shall analyse in detail the imaging properties of this kind of microscopy.

In Chapter 3 we examined the total power in the image of a single point object. We showed that it was independent of defocus in a conventional microscope, but fell off sharply in a confocal system (Fig. 3.29)—so strongly, in fact, that the integral of the power over all axial positions converges. For the extended focus method, then, we elect to integrate in u from $-\infty$ to $+\infty$.

The image of a point object in extended focus is thus

$$I(v) = \int_{-\infty}^{+\infty} [C^2(u, v) + S^2(u, v)]^2 \, du \qquad (5.2)$$

where C and S are defined in equation (2.41).

This is plotted in Fig. 5.9, normalised to unity on the axis, together with the point images for conventional and confocal systems. It is seen that the extended focus image, although very slightly broader than the confocal image, is still sharper than the conventional image. The outer rings are also weaker than in the conventional microscope, and in fact the intensity decreases monotonically with radius.

If axial scanning and integration are used with a conventional microscope, the intensity for a point object may be written

$$I(v) = 4 \int_{-\infty}^{+\infty} \int_0^1 \int_0^1 P(\rho)P^*(\rho') \exp\frac{ja}{2} (\rho^2 - \rho'^2)$$

$$\times J_0(v\rho)J_0(v\rho')\rho\rho' \, d\rho \, d\rho' \, du. \qquad (5.3)$$

Performing the integral in u we obtain

$$I(v) = 4 \int_0^1 \int_0^1 P(\rho)P^*(\rho')J_0(v\rho)J_0(v\rho')\,\delta(\rho^2 - \rho'^2)\rho\rho'\,d\rho\,d\rho' \quad (5.4)$$

where δ is the Dirac delta function. Making the substitutions

$$\left.\begin{array}{l} \rho^2 = p \\ \rho'^2 = p' \end{array}\right\} \quad (5.5)$$

this can be written

$$I(v) = \int_0^1 \int_0^1 P(p)P^*(p')J_0(v\sqrt{p})J_0(v\sqrt{p'})\,\delta(p - p')\,dp\,dp' \quad (5.6)$$

$$= \int_0^1 P(p)P^*(p)J_0^2(v\sqrt{p})\,dp. \quad (5.7)$$

Somewhat surprisingly, this is independent of aberrations. For an unshaded circular pupil we obtain

$$I(v) = 2 \int_0^1 J_0^2(v\rho)\rho\,d\rho \quad (5.8)$$

$$= \frac{2}{v^2} \int_0^v J_0^2(t)t\,dt \quad (5.9)$$

which may be written [5.6]

$$I(v) = J_0^2(v) + J_1^2(v) \quad (5.10)$$

with J_1 a Bessel function of order unity, which is shown in Fig. 5.9. The curve is broad, indicating a poor image, and the total power within a radius v diverges for large v. This infinite incident power results from the integration over an infinite axial distance. In practice, of course, the axial scan and total power must be finite, but the resulting image is still poor if the axial scan is appreciable.

For the case of two closely spaced points, the Rayleigh criterion shows that in reflection, the extended focus case results in a small decrease in two-point resolution as compared to the incoherent conventional microscope, but that the resolution is superior by 7% in transmission [5.3].

We now turn our attention to general objects with spectrum $T(m, n)$. We recall that for a coherent linear space invariant system with coherent transfer function $c(m, n)$, the image intensity is given by the modulus square of the amplitude and may be written as

$$I(x, y) = \int\int\int\int_{-\infty}^{+\infty} c(m, n)c^*(p, q)T(m, n)T^*(p, q)$$

$$\times \exp - 2\pi j\{(m - p)x + (n - q)y\}\, dm\, dn\, dp\, dq. \quad (5.11)$$

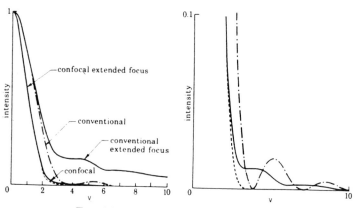

FIG. 5.9. The image of a single point.

Equation (5.11) holds for coherent systems. For a general partially coherent system we must write

$$I(x, y) = \int\int\int\int_{-\infty}^{+\infty} C(m, n; p, q)T(m, n)T^*(p, q)$$

$$\times \exp - 2\pi j\{(m - p)x + (n - q)y\}\, dm\, dn\, dp\, dq \quad (5.12)$$

and in general $C(m, n; p, q)$ does not separate.

We obtain the extended focus confocal image by integrating equation (5.11) over all values of u, remembering that $c(m, n)$ is dependent on the degree of focus. Equation (5.12) thus holds for the extended focus case, in which for reflection

$$C(m, n; p, q) = \int_{-\infty}^{+\infty} c(m, n, u)c^*(p, q, u)\, du. \quad (5.13)$$

Restricting the analysis to line structures such that

$$t(x, y) = t(x) \tag{5.14}$$

only, equation (5.11) reduces to

$$I(x) = \int\int_{-\infty}^{+\infty} C(m; p)T(m)T^*(p) \exp - 2\pi j (m - p)x \, dm \, dp \tag{5.15}$$

where

$$C(m; p) = \int_{-\infty}^{+\infty} c(m, n)c^*(p, u) \, du \tag{5.16}$$

for the extended focus case.

The coherent transfer function of a confocal system is given by the convolution of the defocused pupil functions of the illuminating and collecting lenses (equation 3.42). Thus for equal circular pupils in reflection mode, the transfer function is given by the integral

$$c(m, u) = \int\int_S \exp \left(\tfrac{1}{2}ju\rho_1^2\right) \exp \left(\tfrac{1}{2}ju\rho_2^2\right) \, dS \tag{5.17}$$

over the shaded S (Fig. 5.10).

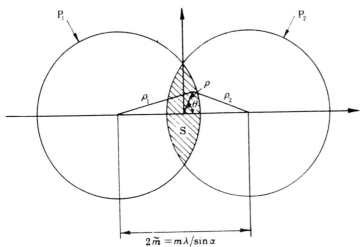

$$2\tilde{m} = m\lambda/\sin\alpha$$

FIG. 5.10. Area of integration for the defocused transfer function of a confocal microscope.

Introducing a normalised spatial frequency

$$\tilde{m} = \frac{m\lambda}{2 \sin \alpha} \tag{5.18}$$

yields

$$\rho_1^2 = \rho^2 + \tilde{m}^2 + 2\rho\tilde{m} \sin \theta,$$
$$\rho_2^2 = \rho^2 + \tilde{m}^2 - 2\rho\tilde{m} \sin \theta. \tag{5.19}$$

Thus

$$\tfrac{1}{2}(\rho_1^2 + \rho_2^2) = \rho^2 + \tilde{m}^2 \tag{5.20}$$

is independent of θ. Substituting in equation (5.17) and normalising so that

$$c(m = 0, u = 0) = 1 \tag{5.21}$$

one obtains

$$c(m, u) = \exp ju\tilde{m}^2 \left\{ 2 \int_0^{1-\tilde{m}} \exp (ju\rho^2)\rho \, d\rho + \frac{4}{\pi} \int_{1-\tilde{m}}^{\sqrt{(1-\tilde{m}^2)}} \right.$$
$$\left. \times \sin^{-1} \left(\frac{1 - \tilde{m}^2 - \rho^2}{2\rho\tilde{m}} \right) \exp (ju\rho^2)\rho \, d\rho \right\} \tag{5.22}$$

of for the real and imaginary parts

$$\left. \begin{aligned} c_{\text{Re}}(m, u) = 2 \int_0^{1-\tilde{m}} \cos u(\tilde{m}^2 + \rho^2)\rho \, d\rho + \frac{4}{\pi} \int_{1-\tilde{m}}^{\sqrt{(1-\tilde{m}^2)}} \\ \times \sin^{-1} \left(\frac{1 - \tilde{m}^2 - \rho^2}{2\rho\tilde{m}} \right) \cos u(\tilde{m}^2 + \rho^2)\rho \, d\rho \\ c_{\text{Im}}(m, u) = 2 \int_0^{1-\tilde{m}} \sin u(\tilde{m}^2 + \rho^2)\rho \, d\rho + \frac{4}{\pi} \int_{1-\tilde{m}}^{\sqrt{(1-\tilde{m}^2)}} \\ \times \sin^{-1} \left(\frac{1 - \tilde{m}^2 - \rho^2}{2\rho\tilde{m}} \right) \sin u(\tilde{m}^2 + \rho^2)\rho \, d\rho \end{aligned} \right\} . \tag{5.23}$$

In each case the first integral can be performed analytically. This coherent transfer function has been evaluated numerically, and its real and imaginary parts are shown in Fig. 5.11. For $m = 0$ equations (5.23) may be evaluated numerically, yielding

$$c_{Re}(m = 0, u) = \frac{\sin u}{u}$$

$$\left.\begin{array}{c}\\ \\ \\ \\ \\ \end{array}\right\} \qquad (5.24)$$

$$c_{Im}(m = 0, u) = \frac{1 - \cos u}{u}$$

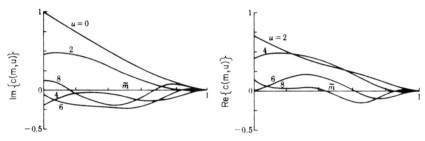

FIG. 5.11. Real and imaginary parts of the defocused transfer function of a confocal microscope.

The coherent transfer function when perfectly focused is, of course, real; and equation (5.23) reduces to the usual expression for the convolution between two circles (equation 3.42).

We now specialise to an object of the form

$$t(x) = \exp (b \cos 2\pi v x) \qquad (5.25)$$

where b is constant and in general complex to allow for both amplitude and phase variations. If the modulus of b is small compared with unity we may, as in Chapter 4, use equation (5.15) to write the image intensity with defocus of this weak object as

$$I(x) = C(0; 0, u) + 2 \, \text{Re} \, (\{b^*C(v; 0, u)\} \cos 2\pi v x \qquad (5.26)$$

or for a coherent microscope

$$I(x) = c(0, u)c^*(0, u) + 2 \, \text{Re} \, \{b^*c(v, 0)c^*(0, u)\} \cos 2\pi v x. \qquad (5.27)$$

The real part of $c(v; 0)$ results in the imaging of amplitude variations, and the imaginary part of $c(v; 0)$ in the imaging of phase structure. Thus for the confocal microscope in reflection we have

$$
\left.\begin{aligned}
C_{\text{Re}}(v; 0, u) &= \left(\frac{\sin u}{u}\right) c_{\text{Re}}(v, u) + \left(\frac{1 - \cos u}{u}\right) c_{\text{Im}}(v, u) \\
C_{\text{Im}}(v; 0, u) &= \left(\frac{\sin u}{u}\right) c_{\text{Im}}(v, u) - \left(\frac{1 - \cos u}{u}\right) c_{\text{Re}}(v, u)
\end{aligned}\right\} . \quad (5.28)
$$

The real and imaginary parts of this weak object transfer function are shown in Fig. 5.12. It is seen that the real part is well behaved for values of u less than about six, by which time the magnitude of the real part is small anyway. The imaginary part, on the other hand, has a magnitude as large as 0·1

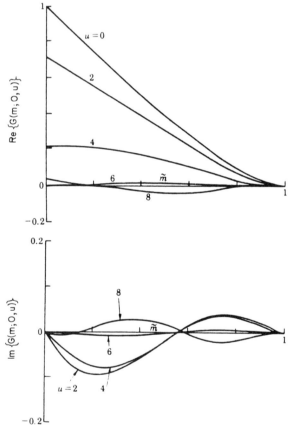

FIG. 5.12. Real and imaginary parts of the defocused weak object transfer function for a confocal microscope.

around $u = 2$, which results in the image of phase variations, perhaps caused by specimen height variations. There are two completely different mechanisms whereby height variations may be imaged in confocal microscopy. In the first, the small variations in height are imaged by the transfer function, assumed constant over the corresponding part of the image. For example, if the transfer function is defocused by the presence of a constant displacement from the focal plane, small variations in height about this constant displacement are imaged by its imaginary part. In the second mechanism, the absolute height variations are imaged by the optical sectioning property. It is difficult to develop a theory which adequately incorporates both these effects, although it is apparent that the first mechanism results in contrast which is independent of numerical aperture, while the second becomes increasingly more important at high numerical apertures.

If we now return to consider the transfer function of the extended focus confocal microscope, we have from equation (5.16), for the weak object transfer function

$$C(m; p) = \int_{-\infty}^{+\infty} \{c_{Re}(m, u)c_{Re}(p, u) + c_{Im}(m, u)c_{Im}(p, u)\} \, du \qquad (5.29)$$

that is, it is wholly real. Here we have made use of the relationship

$$\left.\begin{array}{l} c_{Re}(m, -u) = c_{Re}(m, u) \\ c_{Im}(m, -u) = -c_{Im}(m, u) \end{array}\right\}. \qquad (5.30)$$

Furthermore the relationships

$$\left.\begin{array}{l} c_{Re}(-m, u) = c_{Re}(m, u) \\ c_{Im}(-m, u) = c_{Im}(m, u) \end{array}\right\} \qquad (5.31)$$

result in the symmetry conditions

$$C(\pm m; \pm p) = C(\pm p; \pm m). \qquad (5.32)$$

The weak object transfer function may be written

$$C(m; 0) = \int_{-\infty}^{+\infty} c(m, u)c^*(0, u) \, du \qquad (5.33)$$

where

$$c^*(0, u) = \frac{j}{u}(\exp - ju - 1). \qquad (5.34)$$

Using equation (5.22) and evaluating the integral in u first we have

$$F(\tilde{m}, \rho) = \int_{-\infty}^{+\infty} \frac{j}{u} (\exp - ju - 1) \exp ju(\tilde{m}^2 + \rho^2) \, du \qquad (5.35)$$

$$= \pi[\text{sgn}\, (\tilde{m}^2 + \rho^2) - \text{sgn}\, (\tilde{m}^2 + \rho^2 - 1)] \qquad (5.36)$$

where the sgn function is plus one for positive argument and minus one for negative argument, and the integral

$$\int_{-\infty}^{+\infty} \frac{\exp jax}{x} \, dx = \pi \, \text{sgn}\, a \qquad (5.37)$$

has been used. Thus

$$C(m; 0) = 2 \int_{0}^{1-\tilde{m}} F(\tilde{m}, \rho)\rho \, d\rho + \frac{4}{\pi} \int_{1-\tilde{m}}^{\sqrt{(1-\tilde{m}2)}}$$

$$\times \sin^{-1} \left(\frac{1 - \tilde{m}^2 - \rho^2}{2\tilde{m}\rho} \right) F(\tilde{m}, \rho)\rho \, d\rho \qquad (5.38)$$

$$= C(m; 0, n = 0). \qquad (5.39)$$

The weak object transfer function for the extended focus microscope is thus purely real and identical to that for the confocal microscope (and also the conventional microscope with equal pupils).

The transfer function $C(m; m)$ for the extended focus microscope is also purely real, and given by

$$C(m, m) = \int_{-\infty}^{+\infty} \{c_{\text{Re}}^2(m, u) + c_{\text{Im}}^2(m, u)\} \, du \qquad (5.40)$$

which is shown in Fig. 5.13. It is seen that this is a smooth, monotonically decreasing, well-behaved function which falls off more quickly than the conventional microscope with equal pupils, but more slowly than the confocal microscope.

The transfer function theory which has been presented here has been derived for objects which do not depart appreciably from one plane. In practice, if the object height is slowly varying such that the transfer function may be assumed spatially invariant over each patch of the object, the results may also be applied. In general, if the object height is allowed to vary substantially within the region where the point spread function is appreciable, the method breaks down as the transfer function is not spatially invariant; that is, it is not independent of the object. As we explained earlier, this restriction is more severe at higher numerical apertures.

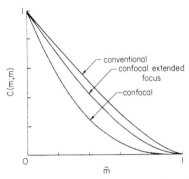

FIG. 5.13. The transfer function $C(m, n, u)$ for focused confocal, conventional and extended focus confocal imaging modes. This corresponds to the image of a perfect reflector at an angle $\sin^{-1}(m\lambda/2)$ to the optic axis.

5.5 Conclusions

The introduction of axial scanning into a confocal microscope has resulted in a technique capable of dramatically extending the depth of field in optical microscopy. The resultant images bear some similarities to those produced in the scanning electron microscope, but do not exhibit the shadowing effects, as the illuminating and detection optics are coaxial. These images are also similar to confocal images in many respects, but do not, of course, exhibit contrast from variations in absolute phase. The confocal microscope requires that the object be scanned while retaining the axial position with high precision as otherwise intensity fluctuations appear in the image. In extended focus microscopy, however, this requirement is relaxed so that the method is less affected by external vibrations.

References

[5.1] D. K. Hamilton, T. Wilson and C. J. R. Sheppard (1981). *Opt. Lett.* **6**, 625.
[5.2] T. Wilson and D. K. Hamilton (1982). *J. Microsc.* **128**, Pt. 2, 139.
[5.3] C. J. R. Sheppard, D. K. Hamilton and I. J. Cox (1983). *Proc. R. Soc. (Lond.)* **A387**, 171.
[5.4] D. K. Hamilton and T. Wilson (1982). *J. Appl. Phys.* **53**, 5321.
[5.5] D. K. Hamilton and T. Wilson (1982). *Appl. Phys.* **B27**, 211.
[5.6] M. Abromowitz and I. A. Stegun (1965). "Handbook of Mathematical Functions". Dover, New York.

Chapter 6

Super-Resolution in Microscopy

6.1 Introduction

The scanning optical microscope has a number of properties which make it particularly suitable for super-resolving methods. Super-resolution can be attained because the image is built up by scanning and the confocal system is a particular example of this application. Because we obtain the image directly in an electronic form, we can measure the image with a high signal-to-noise ratio, particularly if the object is scanned through an unchanging optical system. As a result, the confocal microscope is well suited to methods of digital signal processing.

There are many different possible definitions of resolution. First, the simplest comparison between systems is in terms of the single point response. According to this criterion the confocal system is superior to conventional microscopes. As a resolution criterion, however, it is not really satisfactory, as merely raising the image to some power greater than unity would sharpen up the single point response. This may indeed result in a visually sharper image, but is not really fundamental, because the information content is not increased.

One of the most widely used resolution criteria is the generalised Rayleigh criterion for two-point objects. This criterion also suffers, to a limited degree, from the same objection. We shall refer to two-point resolution greater than that given by the generalised Rayleigh criterion as ultra-resolution. Ultra-resolution may be achieved by apodising the lenses, that is by adjusting the modulus of the pupil function as the radial coordinate varies. A particular example of this is the use of an annular pupil, as discussed earlier.

A more fundamental view of resolution is that determined by the spatial frequency response of the system. But even here there are some problems in making comparisons. A fully coherent or incoherent system is completely specified by a transfer function which will have a definite cut-off frequency. If the pupil is apodised in order to improve the response at high frequencies, the cut-off remains unchanged. This is an improvement in resolution analogous to that described earlier with reference to the two-point resolution: we may call this ultra-resolution. Ultra-resolution results in an improvement in resolution which could be achieved, in principle, by subsequent electronic processing of the conventional image. On the other hand, if the spatial frequency is actually increased, we are transmitting extra information which could in no way be restored to the image by subsequent processing of a conventional kind. We therefore take as our definition of super-resolution the increase in the spatial frequency cut-off beyond that obtainable conventionally.

There are still difficulties in comparing coherent with incoherent systems. Is an incoherent system superior to a coherent one? The incoherent transfer function has twice the spatial frequency cut-off, but the band of spatial frequencies in an intensity object is twice as wide as in an amplitude object. Overall, the incoherent system is superior, as we may see by comparing the region of non-zero transfer function $C(m; p)$: the incoherent transfer function completely includes the coherent one. The confocal microscope has a spatial frequency cut-off equal to that of an incoherent system. However, if we compare the regions where $C(m; p)$ is non-zero, we see that the confocal system passes some extra-high spatial frequencies *in the image* as given by the second and fourth quadrants, whereas the incoherent system passes some extra low spatial frequency components as given by the first and third quadrants. From this we infer that the confocal system is superior.

Compared with a conventional coherent system, the confocal one has twice the spatial frequency cut-off, and may therefore be termed super-resolution. In order to define super-resolution we choose to compare a coherent system with a coherent system or an incoherent one with another incoherent one. We must also now distinguish between two types of super-resolution. Some methods result in obtaining an increase in spatial frequency cut-off for a given pupil, but at the same time limit the aperture of the pupil which may be employed. Relative to a given pupil they give a super-resolution, but the cut-off cannot be increased above that which could, in principle, be achieved by increasing the numerical aperture. Such systems we shall refer to as giving relaxed super-resolution. On the other hand, methods which give an increase in cut-off frequency even at the highest numerical apertures we shall call strict super-resolution. The confocal system may be used at the highest numerical apertures and hence achieves strict super-resolution, although of a fairly limited amount.

We now move on to discuss various practical image processing and super-resolution schemes.

6.2   Digital image processing

In order to perform any processing on an image, it is first necessary to convert it into a suitable form for computer processing. Images from a conventional microscope are usually recorded by using a television camera to convert the image into an electrical signal which can be subsequently digitised. The scanning optical microscope, on the other hand, must scan the object in order to produce an image. Thus the image may be digitised very simply.

In order that an image may be processed accurately and effectively, it is necessary to know the image formation properties of the particular system, i.e. whether the system is coherent or incoherent, the transfer function of the imaging system, etc. In many applications such information is unknown, and it is therefore necessary to estimate the imaging properties. At other times the transfer function may be known, but is so complex that simplifying approximations are applied: in optical microscopy, for example, incoherent imaging may be assumed even though for high resolution the imaging is necessarily partially coherent.

The confocal microscope, however, does not suffer from this drawback, and apart from its purely coherent imaging system it has several desirable features for digital image processing. These include its improved resolution over the conventional microscope using the same aperture and wavelength. Its high precision mechanical scanning allows highly accurate measurements to be made, and also results in space invariant imaging which reduces the effects of aberrations of the lenses, as no off-axis imaging is involved. Further high signal-to-noise ratios, e.g. up to 1000, are attainable. The depth discrimination property also makes range information of the object available, as we can measure the relative distance of its surface from the observer.

A typical system is shown in Fig. 6.1. This consists of a scanning optical microscope controlled by a microcomputer which is attached to a framestore for image display, a terminal, and a floppy disc unit for data and program storage. The microcomputer controls the mechanical scanning and hence the data aquisition rate. It is also useful for automatic focusing, and, of course, for digital image processing.

A 512×512 picture element (pixel) framestore is often chosen [6.1] as it allows a TV quality image to be displayed. Images from framestores with fewer pixels can suffer from pixellation; that is, the condition in which the individual pixels are clearly visible. If the object is sampled at $0\cdot1\text{-}\mu\text{m}$ spacings, a field of view of $51\cdot2\,\mu\text{m}$ is then available. Of course, this field of view may be extended by increasing the sample spacing, albeit at the expense of reduced resolution. Each pixel is represented by eight bits so that 256 grey levels may be recorded and displayed. This may at first seem rather excessive, as the human eye can perceive no more than 64 distinct grey levels at once. However, this byte† format is especially suitable for manipulation by computers, and prevents noise introduced by digital image processing from reducing the final image quality to an unacceptable level.

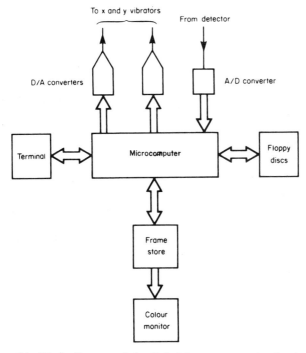

FIG. 6.1. Block diagram of the digital image processing hardware.

Digital image processing is computationally severe, primarily because of the large data set, e.g. $\frac{1}{4}$ Mbyte for a $512 \times 512 \times 8$ bit image. The microcomputer must be capable of addressing this quantity of data and performing simple processing tasks within an acceptable period of time. Eight-bit microcomputers are therefore unacceptable. A 16-bit machine offers a good compromise between cost and performance.

6.2.1 The auto-focus image

If the object in a confocal microscope is scanned axially, the focal position is easily recognised, as it corresponds to the point of maximum detected signal.

† 1 byte = 8 bits, 2^{10} bytes = 1 kbytes, 2^{20} bytes = 1 Mbyte.

This effect has been discussed in detail in the previous chapter. If instead of measuring the axial displacement required to bring a particular object point into focus, as we did for surface profilometry, we formed an image at which each point has been shifted into the focal plane before the intensity is recorded, we would generate a high resolution auto-focus image eliminating the depth discrimination effect. Figure 6.2 shows such an auto-focused image of an integrated circuit, and Fig. 6.3 the conventional image for comparison.

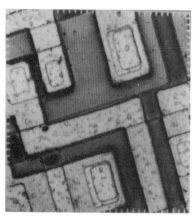 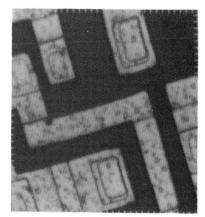

FIG. 6.2. An auto-focused confocal image of an integrated circuit.

FIG. 6.3. Conventional image of an integrated circuit. Each division represents 4 μm.

Notice the improved resolution in the auto-focus image, and also that the aluminium stripes are of uniform intensity. This image could, of course, also be produced by the analogue techniques of Chapter 5.

6.2.2 Pseudo-colour image

A pseudo-colour image is one in which each grey level in the image is mapped to an individual colour. Since the eye can perceive many more colours than grey levels, pseudo-colour may be considered as a further contrast enhancement mechanism. It is also easier to recognise areas of similar intensity in different parts of the field by comparing colours rather than grey levels.

The arbitrariness permitted by forcing a mapping from one to three dimensions is both the advantage and the disadvantage of pseudo-colour. A particular pseudo-colour mapping is therefore not appropriate for all images.

6.2.3 *Contrast enhancement*

Since the image in a scanning optical microscope is always in the form of an electronic signal, contrast enhancement has always been possible. The main advantage of digitising the image in this context is that many more sophisticated contrast enhancement techniques may be employed [6.2].

One technique, which is very useful in the examination of areas of almost uniform brightness, consists of shifting a particular grey level to mid-grey and increasing the slope of the grey level mapping, as shown in Fig. 6.4. Figure

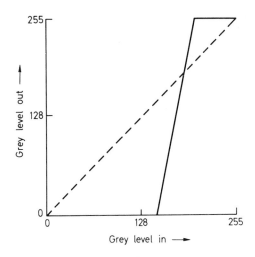

FIG. 6.4. The effect of increasing the slope of the grey level mapping.

6.5 is an example of this technique in which the slope was increased by a factor of three at a mid-grey level corresponding to the aluminium. The algorithm has emphasised the surface texture and edge detail of the aluminium. The technique is also useful in transmission microscopy, where weak modulation objects such as biological specimens may be examined without staining.

The relative distribution of each grey level of the integrated circuit in Fig. 6.3 is shown in Fig. 6.6. Histogram equalisation or flattening attempts to alter the grey level histogram so that each grey level is equally likely. This is achieved by merging neighbouring levels. Thus, the final image contains fewer levels than the original; however, since the eye can only distinguish 64 distinct grey levels this reduction is permissible. Figure 6.7 is the image of the integrated circuit after histogram equalisation. A significant increase in perceived image detail is evident, which is especially striking when we consider that the original image would normally be considered of high contrast.

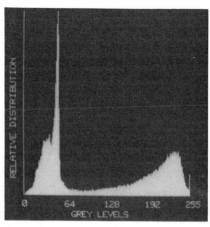

FIG. 6.5. An image where the grey level slope has been increased by a factor of three at the mid grey level corresponding to aluminium.

FIG. 6.6. Grey level histogram of the integrated circuit of Fig. 6.3.

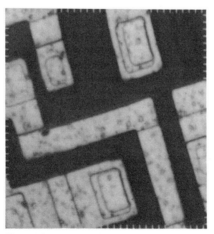

FIG. 6.7. Histogram equalised image of the integrated circuit.

FIG. 6.8. Edge enhanced version of the auto-focused image of Fig. 6.2.

6.2.4 *Edge enhancement*

Edge enhancement or sharpening techniques are designed to increase the visibility of low contrast edges and often lead to an increase in perception of detail. A variety of such techniques are available, the particular choice being governed by the imaging properties of the optical system which produced the image.

A good review of various general techniques is given by Pratt [6.3]. These range from spectral decomposition techniques, which modify the spatial frequency spectrum in the image, to statistical differencing, which essentially divides each pixel value by its measured standard deviation computed over some neighbourhood near the pixel. One may also achieve edge enhancement by convolving the image with a suitable filter. A problem with applying such techniques to scanning optical microscopy is that they assume a linear, i.e. incoherent, imaging system. However this does not prevent such techniques being usefully employed on microscope images. As an example, Fig. 6.8 shows an edge enhanced image of the auto-focused image shown in Fig. 6.2 after convolution with a Laplacian filter [6.1, 6.2]. The Laplacian filter is used in incoherent signal processing to add a small amount of second derivative to the original image. However, as a consequence of the coherence of the confocal imaging, we add a small amount of first derivative to the original image as well. Nonetheless, the improvement in subjective image quality of the auto-focused edge enhanced imaged compared to the conventional image is remarkable, especially considering the simplicity of the digital processing techniques involved. Edge enhancement increases the response of high spatial frequencies without extending the cut-off frequency, and hence is an example of ultra-resolution.

6.2.5 *The range image and stereo pairs*

While recording the auto-focus image, it is also possible simultaneously to record the relative variations in height of the object in order to display a range image as shown in Fig. 6.9. Here, the lighter the point is, the closer it is to the observer.

Such an image in association with the auto-focus image allows the observer to acquire an understanding of the three-dimensional structure of the object which would not be possible otherwise.

The data available in the auto-focus and range images allows a stereoscopic image pair to be generated by computer. The theory behind such an image pair is relatively simple. Figure 6.10 illustrates the image-forming process, from which it is easily shown that the x-coordinates for the left and right eyes, x_l, x_r, respectively are given by

$$x_l = x_s + \frac{DS}{2z}$$

$$x_r = x_s - \frac{DS}{2z}$$

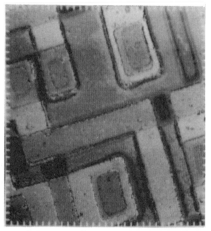

FIG. 6.9. A range image of the integrated circuit. Lighter areas correspond to points closer to the observer.

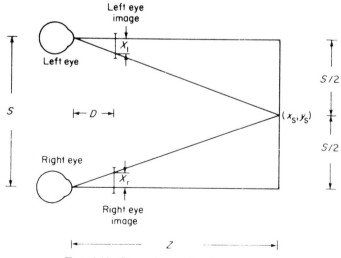

FIG. 6.10. Stereo image forming process.

and

$$y_l = y_r$$

where the stereoscopic pair is positioned a distance D in front of the viewer's eyes, S is the interocular distance, z is the distance of the point from the observer and x_s, y_s are the x, y coordinates for monocular vision. Figure 6.11 shows a stereoscopic image pair, which may be viewed using conventional techniques.

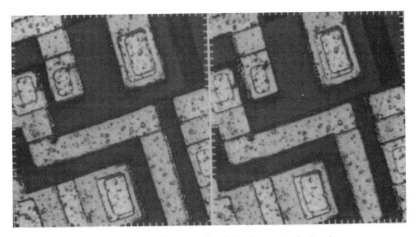

FIG. 6.11. Stereo image pair of the integrated circuit.

It should be noted that, in a confocal system, the digital processing required to examine the three-dimensional qualities of an object is considerably less than in a conventional system, where auto-focusing is more complex and image processing must be applied in order to remove the blurred out-of-focus planes of a specimen [6.4].

6.2.6 General comments

We have just seen that by applying very simple digital processing techniques to confocal microscope signals, we may generate images which depend on a great variety of object properties. The imaging properties of the confocal microscope are particularly suited to digital processing, and one may consider applying more sophisticated techniques such as analytic continuation to vastly increase the resolution in the images.

The discussion so far has concentrated on non-interference images, but great advantages would result from processing the images from scanning interference microscopes. We have shown in Chapter 4 that a confocal interference microscope may simultaneously produce images of the real and imaginary parts of the object amplitude transmittance in a linear fashion, even for strong objects. These two signals may then be processed to yield final images depending only on the modulus and phase of the object amplitude transmittance.

We have made no mention of noise in any of our previous discussion. Noise may arise from many sources, such as electrical detector noise or channel error in the digitised signal, which may be minimised by classical statistical filtering techniques. However, in practice, the image noise tends to appear at discrete isolated pixel locations which are *not* spatially correlated. The noisy pixels often appear markedly different from their neighbours and hence "outrange" noise cleaning methods may be employed. Noise also tends to have high spatial frequency components, which allows the use of spatial filtering. We should finally mention that many processing techniques enhance the noise as well as the object property of interest. For this reason care should be taken to produce an image as noise free as possible before processing.

6.3   Multiple traversing of the object

The improvement in resolution and imaging performance of the confocal microscope results from the fact that two lenses are taking part in the imaging. One is tempted to ask, therefore, whether a further increase in resolution would result by using a higher number of lenses. A detailed analysis of this case, where the radiation traverses the object n times, has been given in reference [6.5], and the basic principle of the scheme is shown in Fig. 6.12. The object is illuminated by a plane wave, and light from the object point A is reflected back from a mirror onto the object point B. Suppose that the object consists of an opaque screen with a small hole at A. Then, after reflection, the hole is illuminated with an amplitude $h(2x)$, where h is the impulse response of the mirror. The presence of the factor two in the argument of the impulse response results in a very sharp image of the single point object. However, after reflection in the mirror, the object is illuminated by an amplitude equal to its own mirror image about the optic axis—we are in fact imaging the autoconvolution of the object.

Suppose that instead of a single point the object consisted of two points. The image would then consist of the images of the two points separately plus an interference term, resulting in a bright spot midway between the two points. This latter spot is detected when the object is symmetrically placed about the optic axis and has twice the amplitude of the spots at the two points. A microscope of this sort would clearly be useless. The full analysis [6.5], where the object is illuminated not with a plane wave but with a focused spot, bears out these general conclusions. We find that the image of a single point for two passes through the object is 2·4 times as sharp as that in a conventional microscope if point source and detector are used. Moreover a "ghost" image appears between the images of a pair of point objects. It may be made quite small in the confocal arrangement, but the overall improvement in Rayleigh criterion is small compared to the value for a conventional incoherent microscope. The straight-edge response is also slightly improved, but fringing results if an annular lens is used. In the double pass microscope, amplitude behaves similarly to intensities in a partially coherent system.

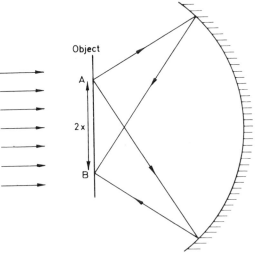

FIG. 6.12. Principle of operation of the double pass microscope.

The overall conclusion of a fully analysis of multiple pass microscopy is that although very slight improvements are possible, the optimum number of passes is one, and that the confocal arrangement is preferred. We should also comment on the similarities between this system and the resonant microscope of Chapter 4. In that case, however, instead of considering the image formed by a beam which traverses the object a given number of times, we must consider the sum of all the beams which traverse the object any number of times.

6.4 Super-resolution by use of a saturable absorber

This technique depends on coating the object with a substance which has a resonant absorption at the wavelength of operation of the microscope. Such absorption becomes non-linear at high enough light intensity, and saturates.

This means that the higher the intensity, the lower the loss; so that a probe with a gaussian intensity distribution will experience less loss near the axis than away from it. As a result, the light that penetrates through such a layer (assumed to be thin in comparison to the wavelength) has a distribution narrower than the incident intensity distribution. If the incident distribution were diffraction limited, super-resolution would result.

6.5 Super-resolution aperture scanning microscopy

As we have already mentioned the problem of super-resolution is greatly simplified by resorting to a scanning technique, as here we can illuminate one resolution element of the object at a time. The method we are about to describe involves illuminating the object through a small aperture of radius r_0, $(r_0 \ll \lambda)$ in a thin diaphragm. Since the hole is small compared to the wavelength, the fields are static and may be approximated by a pair of magnetic and electric dipoles. Hence the fields at the object (which must, of course, be placed very close, of the order of r_0, to the aperture) and the energy stored in the intervening space, depend on the electric and magnetic permittivities and conductivity distribution in the object. These fields contain the information from which we can construct the image.

In order to demonstrate the feasibility of building a microscope which detects these fields, Ash and Nicholls [6.6] built a microwave analogue (Fig. 6.13). Here the object is illuminated through a hole in an open resonator. The dipole field above the object also contains the necessary information to form

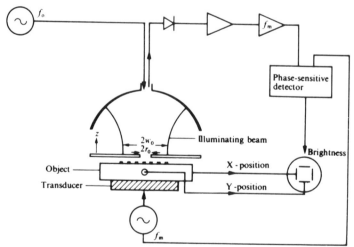

FIG. 6.13. Schematic arrangement of the super-resolving microscope (from E. A. Ash and G. Nicholls (1972). *Nature, Lond.* (June 30), 237).

an image, so that the resonator itself may be used to collect the image signal. However, changes in the reflected signal may occur for a variety of other reasons. Therefore the object is scanned at a frequency f_m and a phase sensitive detector is used to ensure that only information at frequency f_m is displayed on the screen.

Ash and Nicholls reported a resolution of $\lambda/60$ in their first paper [6.6], and subsequently Husain and Ash [6.7] demonstrated a line scan resolution capability of $\lambda/200$. The technique may be used in principle in optical microscopy. There is no particular difficulty in generating sub-wavelength small holes, but the practical problems of scanning the object within a distance r_0 of the aperture are considerable. One would also be limited to examining specimens which either are or could be prepared with a flatness significantly better than an optical wavelength.

6.6   Scanning incoherent confocal fluorescence microscopy

The confocal scanning microscope is a coherent imaging system with twice the spatial frequency bandwidth of the conventional instrument. This super-resolution may be thought of as being due to the restriction of the image field by the detection system [6.8]. It is interesting to speculate on the further improvement in resolution which would be obtained with an incoherent confocal microscope [6.9]. This would have a bandwidth four times larger than the conventional coherent microscope. Such an instrument may be constructed by modifying the confocal microscope so that the radiation leaving the object is incoherent. This incoherent intensity field is then imaged by the second lens in the usual manner. This modification is difficult to achieve by using a subsidiary diffuser in the object plane to remove the spatial coherence, as an efficient scatterer necessarily has granularity which will degrade the image. However fluorescence in the object [6.10] may be used to destroy the coherence, as the fluorescent field is proportional to the intensity of the incident radiation.

Consider a scanning microscope in which the radiation is focused onto the object, and the fluorescent light emitted by the object is focused onto a detector with an arbitrary variation in sensitivity. If we are examining a fluorescent object, the field just beyond the object is proportional to $|h_1|^2 f$ where h_1 is the amplitude point response of the first lens and f represents the spatial distribution of the fluorescent generation. We may use our usual methods of Fourier optics to write the intensity image in the form

$$I(x_s) = \int\int_{-\infty}^{+\infty} \left| h_1\left(\frac{x_1}{\lambda_1 d}\right) h_2\left(\frac{x_2 - x_1}{\lambda_2 d}\right) \right|^2 f(x_s - x_1) D(x_2) \, dx_1 \, dx_2 \quad (6.1)$$

where $D(x_2)$ represents the detector sensitivity, λ_1 and λ_2 are the wavelengths of the incident and fluorescent radiation respectively, x_s represents the scan position and d is the distance of the lenses from the object.

If we now introduce the spectrum of the fluorescent generation

$$F(m) = \int_{-\infty}^{+\infty} f(\xi) \exp - 2\pi jm\xi \, d\xi. \tag{6.2}$$

We can write the image in the usual form

$$I(x_s) = \int_{-\infty}^{+\infty} c(m)F(m) \exp 2\pi jmx_s \, dm \tag{6.3}$$

where

$$c(m) = \int_{-\infty}^{+\infty}\int \left| h_1\left(\frac{x_1}{\lambda_1 d}\right) h_2\left(\frac{x_2 - x_1}{\lambda_2 d}\right) \right|^2 \exp - 2\pi jmx_1 D(x_2) \, dx_1 \, dx_2. \tag{6.4}$$

The transfer function for the conventional fluorescent microscope which employs a large area detector is given by

$$c(m) = \left\{ \int_{-\infty}^{+\infty} |P_2(\lambda_2 dm)|^2 dm \right\} \otimes (P_1 \otimes P_1^*)(\lambda_1 dm). \tag{6.5}$$

We see that such a microscope has the same spatial frequency bandwidth as a conventional incoherent instrument operating at the *primary* wavelength. This is different from the conventional fluorescent microscope operating at the *fluorescent* wavelength, and hence the resolution is better in practice in the scanning case.

The more interesting case, however, is the confocal instrument, for which the transfer function is

$$c(m) = [(P_1 \otimes P_1^*)(\lambda_1 dm)] \otimes [(P_2 \otimes P_2^*)(\lambda_2 dm)]. \tag{6.6}$$

From this we can see that the cut-off is equal to the sum of the cut-offs for conventional incoherent instruments operating at primary and fluorescent wavelengths. In the limiting case, when $\lambda_2 = \lambda_1$, the cut-off is twice that of the conventional incoherent instrument; that is, four times that of the conventional coherent microscope. The transfer function is shown in Fig. 6.14 for a confocal microscope with two equal circular lenses for a range of fluorescent wavelengths ranging from $\lambda_1 \leqslant \lambda_2 < \infty$. The cut-off frequency is proportional to $(1/\lambda_1 + 1/\lambda_2)$, which becomes smaller as λ_2 increases; a typical value being $\lambda_2 \approx 1{\cdot}5\lambda_1$ for fluorescent microscopy.

If we consider the two-point resolution in terms of the Rayleigh criterion, we find [6.9] that the confocal incoherent system can resolve points a factor of $1{\cdot}8$ closer than a conventional coherent system, and a factor of $1{\cdot}3$ closer than a conventional incoherent system. The use of annular pupils results in a further improvement.

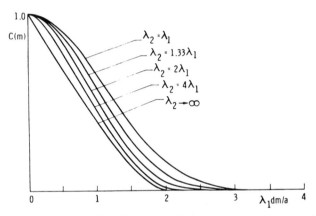

FIG. 6.14. Transfer function for the confocal fluorescent microscope for various fluorescent wavelengths. The spatial frequency axis is normalised by the incident wavelength.

Although our remarks have concentrated on fluorescent microscopy, they apply equally well to any system where the coherence is destroyed in the object plane. This is difficult to achieve with a subsidiary diffuser such as a ground glass screen but a possible approach would be to use subsidiary fluorescence. If a fluorescent (or even cathodoluminescent) material is placed close to the object, the size of the light spot incident on the specimen is limited by the wavelength of the *primary* radiation. The corresponding resolution can be obtained, but with a contrast determined by the wavelength of the *fluorescent* radiation. With a fluorescent wavelength close to that of the primary radiation, the four-fold improvement in spatial frequency bandwidth would be achieved.

References

[6.1] I. J. Cox and C. J. R. Sheppard (1983). *Image and Vision Computing*, **1**, 52.
[6.2] E. I. Hall (1981). "Computer Image Processing and Recognition". Academic Press, New York and London.

[6.3] W. K. Pratt (1978). "Digital Image Processing". Wiley, New York.
[6.4] K. Castleman (1979). "Digital Image Processing". Prentice Hall, Englewood Cliffs, New Jersey.
[6.5] C. J. R. Sheppard and T. Wilson (1980). *Opt. Acta* **27**, 611.
[6.6] E. A. Ash and G. Nicholls (1972). *Nature* **237**, 510.
[6.7] A. Hussain and E. A. Ash (1973). Proceedings 3rd European Microwave Conference, Brussels, September 1973.
[6.8] W. Lukosz (1966). *JOSA* **56**, 1463.
[6.9] I. J. Cox, C. J. R. Sheppard and T. Wilson (1982). *Optik* **60**, 391.
[6.10] C. W. McCutchen (1967). *JOSA* **57**, 1190.

Chapter 7

The Direct View Scanning Microscope

7.1   Introduction

Our discussions in the previous chapters of this book have been concerned with scanning microscopes and in particular the improvements in resolution and imaging which may be obtained in the confocal arrangement. We have already seen that the Type 1 scanning microscope behaves identically to the conventional microscope with the rôles of the two lenses reversed. Thus it seems reasonable to ask if it is possible to modify a conventional instrument so as to obtain "direct view" confocal imaging.

The basic reason for the improvement in the confocal case is that we focus radiation from a point source onto one point of the object, and then, by using a limiting aperture in the detector plane, arrange to collect light from the same portion of the object (if the two lenses are equal) that the first lens illuminates. In this way both lenses contribute equally to the imaging. In a practical instrument we usually elect to scan the object, but in principle the same image would be obtained if instead we scanned the source and detector together.

This immediately suggests that we could build a direct view confocal microscope by having many corresponding confocal points in the source and detector planes such that the object is probed simultaneously by many points of light, but that only light from a particular source point is collected by the corresponding detector point. If we now scan the source and detector, we essentially have many confocal microscopes operating in parallel, which results in a real time image similar to that which would be obtained in a confocal scanning microscope.

We now recall that when light from an extended incoherent source is seen by a lens from such a distance that the maximum path difference between extreme rays is less than say, a quarter of a wavelength, then it has some of the attributes of coherent light. A scanning optical microscope could, in principle, be constructed using such light. This condition is equivalent to ensuring that the size of the source is small compared to the Airy disc of the lens in the plane of the source. It is demonstrable that the amount of light then available is too small to be of practical interest. This is why laser radiation is used in scanning microscopes, because the amount of light available is practically unlimited. It is possible, however, to use a large number of incoherent light sources in parallel so as to obtain adequate light intensity. Thus a direct view microscope can be built using an incandescent tungsten filament or mercury arc lamp.

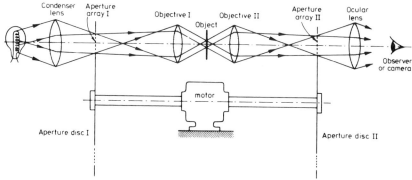

FIG. 7.1. Schematic diagram of the direct view scanning microscope.

A practical version of such a microscope is shown in Fig. 7.1. The transparent object is placed between two microscope objectives I and II in the plane of the image of the aperture array in disc I. Aperture array I is illustrated by light from an incoherent source concentrated by the condenser lens. Lens II images the object *and* the image of array I onto aperture array II in disc II. This real image is then viewed or photographed via the ocular lens, being built up by rotating the discs at such a speed that the image is flicker free. It is also possible to build a reflection form of the microscope which has the advantage of needing only one aperture disc. This is imaged back on itself by the microscope lens, but the available light is then reduced by a beam splitter.

The amount of light reaching the object is reduced below that of a conventional microscope by the ratio of the area of the holes to the total area. As the hole diameter is decreased the image intensity also decreases, but one would expect the resolution to increase. It is therefore important to use the largest possible hole diameter which is consistent with a substantial improvement in the imaging. We will return to this point in more detail in section 7.2.4.

7.2   A unified theory of image formation

The theory of imaging in scanning microscopes with finite incoherent source and detector, which we discussed at the end of Chapter 3, is unfortunately not directly applicable to the direct view microscope. Therefore we now develop a unified theory which is applicable to both scanning, conventional and direct view microscopes. Considering the arrangement of Fig. 7.2, and using the methods of Chapter 3, we can write the intensity at the detector as

$$
I(x_2, x_s) = \int\int\int_{-\infty}^{+\infty} S(x_1 - Mx_s) h_1\left(\frac{x_1/M - x_0}{\lambda d}\right)
$$

$$
\times\, h_1^*\left(\frac{x_1/M - x_0'}{\lambda d}\right) t(x_0) t^*(x_0') h_2\left(\frac{x_2/M - x_0}{\lambda d}\right) h_2^*\left(\frac{x_2/M - x_0}{\lambda d}\right)
$$

$$
\times\, D(x_2 - Mx_s)\, dx_1\, dx_0\, dx_0'. \tag{7.1}
$$

Here again we have restricted ourselves to one dimensional for simplicity.

We can now examine special cases. If both source and detector distributions are unity for all x_1, x_2, the intensity in the x_2 plane is independent of x_s and we obtain

$$
I(x_2) = \int\int\int_{-\infty}^{+\infty} h_1\left(\frac{x_1/M - x_0}{\lambda d}\right) h_1^*\left(\frac{x_1/M - x_0'}{\lambda d}\right) t(x_0) t^*(x_0')
$$

$$
\times\, h_2\left(\frac{x_2/M - x_0}{\lambda d}\right) h_2^*\left(\frac{x_2/M - x_0'}{\lambda d}\right) dx_1\, dx_0\, dx_0' \tag{7.2}
$$

which corresponds to the conventional microscope. The integral in x_1 may be evaluated first, and thus we may again confirm that the aberrations of the first lens are unimportant in conventional microscopes. Equation (7.2), of

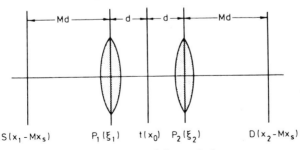

FIG. 7.2. Geometry of the optical system.

course, represents partially coherent imaging, which becomes coherent or incoherent as the numerical aperture of the first lens is made much smaller or larger than that of the second lens.

If we now integrate equation (7.1) over x_2, the intensity is a function of the scan position and the microscope becomes the generalised scanning microscope with partially coherent source and detector which we discussed briefly in Chapter 3.

On the other hand, if we integrate over x_s, we produce the direct view scanning microscope, in which the image is built up in the x_2 plane as the source and detector pupils are scanned. Then

$$I(x_2) = \int\limits_{-\infty}^{+\infty}\!\!\int\int\int S(x_1 - Mx_s) h_1\!\left(\frac{x_1/M - x_0}{\lambda d}\right)$$

$$\times\, h_1^*\!\left(\frac{x_1/M - x_0'}{\lambda d}\right) t(x_0) t^*(x_0') h_2\!\left(\frac{x_2/M - x_0}{\lambda d}\right) h_2^*\!\left(\frac{x_2/M - x_0'}{\lambda d}\right)$$

$$\times\, D(x_2 - Mx_s)\, \mathrm{d}x_1\, \mathrm{d}x_0\, \mathrm{d}x_0'\, \mathrm{d}x_s. \tag{7.3}$$

If both source and detector are point-like, we have

$$I(x_2) = \int\limits_{-\infty}^{+\infty}\!\!\int h_1\!\left(\frac{x_1/M - x_0}{\lambda d}\right) h_1^*\!\left(\frac{x_1/M - x_0'}{\lambda d}\right) t(x_0) t^*(x_0')$$

$$\times\, h_2\!\left(\frac{x_2/M - x_0}{\lambda d}\right) h_2^*\!\left(\frac{x_2/M - x_0'}{\lambda d}\right) \mathrm{d}x_0\, \mathrm{d}x_0' \tag{7.4}$$

which may be written

$$I = |h_1 h_2 \otimes t|^2$$

where $\otimes$ denotes the convolution operation. The image is thus exactly as for the confocal scanning microscope, exhibiting the usual advantages that the resolution is improved and that the side lobes in the point spread function are reduced, enabling apodisation to introduce further resolution improvement.

If both source and detector are large, on the other hand, the microscope becomes a conventional microscope, and performance is that of a Type 1 scanning microscope. It should be noted that the direct view scanning microscope produces an image for any source and detector distribution, unlike both the normal scanning microscope and the conventional microscope.

Returning to the direct view scanning microscope with extended source and detector, the integral in x_s in equation (7.1) may be evaluated first, to give

$$Q(x_1 - x_2) = \int_{-\infty}^{+\infty} S(x_1 - Mx_s)D(x_2 - Mx_s)\,dx_s \qquad (7.5)$$

which is a function of $x_1 - x_2$ and is equal to the convolution of the source and detector distributions if these latter are symmetrical. Thus it may be seen that if either source or detector is large, then imaging is that of a Type 1 scanning microscope regardless of the size of the other. Putting equation (7.5) into equation (7.3), we obtain

$$I(x_2) = \int_{-\infty}^{+\infty}\int Q(x_1 - x_2)h_1\left(\frac{x_1/M - x_0}{\lambda d}\right)$$

$$\times h_1^*\left(\frac{x_1/M - x_0'}{\lambda d}\right)t(x_0)t^*(x_0')h_2\left(\frac{x_2/M - x_0}{\lambda d}\right)$$

$$\times h_2^*\left(\frac{x_2/M - x_0'}{\lambda d}\right)dx_1\,dx_0\,dx_0'. \qquad (7.6)$$

7.2.1 The image of a point object

Let us now consider the image of the simplest object, a single point. Substituting

$$t(x_0) = \delta(x_0) \qquad (7.7)$$

where δ is the Dirac delta function, into the relevant equations, we obtain for the conventional microscope (with or without source)

$$I = h_2 h_2^*. \qquad (7.8)$$

For the scanning microscope with extended source and detector we have

$$I = (h_1 h_1^* \otimes S)(h_2 h_2^* \otimes D) \qquad (7.9)$$

which reduces to

$$I = h_1 h_1^* \qquad (7.10)$$

for the Type 1 scanning microscope and

$$I = h_1 h_1^* h_2 h_2^*$$

for the confocal scanning microscope. For the direct view scanning microscope

$$I = (h_1 h_2^* \otimes Q) h_2 h_2^* \tag{7.11}$$

which, of course, reduces to equation (7.8) if the extent of Q is large, and to equation (7.10) if the extent of Q is small.

7.2.2 Fourier imaging

Following Chapter 3, we may introduce the Fourier transform of the object transmittance. For scanning microscopes we again obtain the general form of the transfer function $C(m; p)$ as in Chapter 3. However, we are particularly interested in the direct view microscope, for which we obtain from equation (7.6) after some routine manipulation

$$C(m; p) = \int\int_{-\infty}^{+\infty} F_Q \left[\frac{\xi_1' - \xi_1}{M\lambda d} \right] P_1(\xi_1) P_1(\xi_1')$$

$$\times P_2(\lambda dm - \xi_1) P_2^*(\lambda dp - \xi_1') \, d\xi_1 \, d\xi_1' \tag{7.12}$$

where F_Q is the Fourier transform of Q.

The geometrical interpretation of this equation is shown in Fig. 7.3, in which we show P_1 and P_2 as unity for $|\xi| < a_{1,2}$ and zero otherwise. For

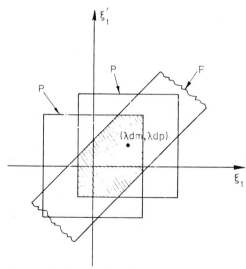

FIG. 7.3. Geometrical evaluation of the transfer function $C(m; p)$.

simplicity we have assumed that F_Q is a simple function which is zero outside a certain range and unity otherwise. The transfer function is equal to the area in the ξ_1, ξ_1' plane which is common to the squares of sides a_1 and a_2 (the latter being displaced by λdm, λdp) and which also lies within the strip F_Q which lies parallel to the line $\xi_1 = \xi_1'$. In practice, of course, the magnitude of F_Q would vary with distance.

We are particularly interested in the value of the transfer function for $p = 0$, as this tells us how weak objects are imaged. If the system is symmetrical, we obtain

$$C(m; 0) = \{[P_1^*(\lambda dm)P_2^*(\lambda dm) \otimes F_Q(m/M)]P_1(\lambda dm)\} \otimes P_2(\lambda dm). \quad (7.13)$$

If we restrict the calculation to the case when P_1 and P_2 are unity for $\lambda dm < a$ and zero otherwise, we see that if F_Q corresponds to either a point source or detector or an infinite source and detector, we obtain the result.

$$C(m; 0) = P(\lambda dm) \otimes P(\lambda dm). \quad (7.14)$$

For weak objects, therefore, a confocal direct view microscope exhibits no improvement in resolution over the Type 1 arrangement, as is also true for the corresponding scanning microscopes.

7.2.3 *The effect of aperture size and distribution in the direct view microscope*

We are now in a position to consider the most suitable size and distribution of apertures in a direct view scanning microscope, so that we obtain most of the improvement of confocal microscopy with the maximum light efficiency. We shall consider two different effects of such aperture size and distribution. First, we shall consider the effect on Fourier imaging, and then the effect on the effective point spread function.

We showed in the previous section that the effect of the hole size does not affect the value of $C(m; 0)$ for aberration free slit pupils. This is also true for $C(m; m)$, the transfer function for difference frequencies which result in zero spatial frequency in the image. By consideration of the form of the transfer function, it is apparent that the maximum effect of the source and detector distributions is on the value $C(m; -m)$, the transfer function for sum frequencies in the image. We therefore examine the effect of the aperture size on $C(m; -m)$, assuming that the apertures are well spaced. Let us assume that the source and detector apertures are of width $2b$, which results in the function F_Q being of the form sinc^2, where the sinc function is defined as

$$\text{sinc}(x) = \frac{\sin \pi x}{\pi x}. \quad (7.15)$$

For confocal operation we require F_Q to be unity over the range of the pupil functions, as shown in Fig. 7.3, and thus a reasonable approximation will result if we take the sinc^2 function as falling to zero at the edge of the lens pupil. This gives a relationship

$$b \lesssim \frac{M\lambda d}{4a} \qquad (7.16)$$

where a is the radius of the lens pupils.

Let us now assume a spacing between adjacent apertures of l, so that if we neglect the size of each aperture, the source (and detector) is given by

$$S(x) = \text{comb}\left(\frac{x}{l}\right) = \sum_{n=-\infty}^{\infty} \delta(x - n). \qquad (7.17)$$

The quantity F_Q is then also a comb function, and if the aperture spacing is such that

$$l < \frac{M\lambda d}{2a} \qquad (7.18)$$

only the central spike lies within the pupil, and imaging is exactly as in a conventional microscope. Calling this separation l_0 we show in Fig. 7.4 the transfer function $C(m; -m)$ for various values of l. It is seen that if l is equal to $3l_0$ we have quite a good approximation to confocal imaging, i.e. we require

$$l \gtrsim \frac{3M\lambda d}{2a}. \qquad (7.19)$$

An interesting feature is that if the separation is less than l_0, the imaging is as for a conventional microscope regardless of the size of the individual holes.

Let us now examine the effect of imaging a single point. Assuming

$$S(x) = \text{rect}\left(\frac{x}{2b}\right) \text{comb}\left(\frac{x}{l}\right) \qquad (7.20)$$

representing an array of apertures, where the function rect is defined as

$$\text{rect}(x) = \begin{cases} 1 - |x|; & |x| < 1 \\ 0; & \text{otherwise} \end{cases} \qquad (7.21)$$

FIG. 7.4. Transfer function $C(m; -m)$ for arrays of point apertures of various spacings, (a) $l = 2l_0$, (b) $l = 3l_0$ and (c) $l = 4l_0$. The transfer function for arrays of slits of width $2b = l_0$ is also shown. Transfer functions for Type 1 and confocal microscopes are plotted by full lines for comparison.

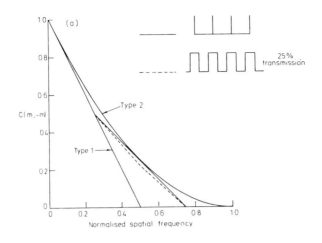

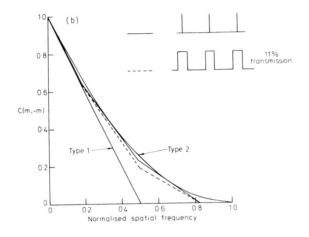

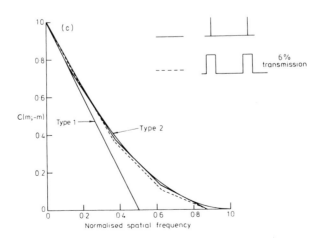

we have

$$Q(x) = \Lambda\left(\frac{x}{2b}\right) \otimes \mathrm{comb}\left(\frac{x}{l}\right) \tag{7.22}$$

with

$$\Lambda(x) = \begin{cases} 1 - |x|; & |x| < 1 \\ 0; & \text{otherwise} \end{cases} \tag{7.23}$$

and therefore

$$I(x) = \left\{h_1^2\left(\frac{x}{M\lambda d}\right) \otimes \Lambda\left(\frac{x}{2b}\right) \otimes \mathrm{comb}\left(\frac{x}{l}\right)\right\}h_2^2\left(\frac{x}{M\lambda d}\right). \tag{7.24}$$

For a slit pupil the first zero of the lens point spread function occurs at a distance of $M\lambda d/2a$, which gives as a condition for b

$$b \ll \frac{M\lambda d}{4a} \tag{7.25}$$

agreeing with equation (7.16). It is advantageous to make the distances between apertures equal to the distance between the origin and a null of the point spread function. This should reduce the amount of light reaching an aperture from a source of which it is not an image, and thus reduce the impairment of the imaging due to the multiplicity of apertures. This gives as a condition for l.

$$l = n\frac{M\lambda d}{2a}, \tag{7.26}$$

n being an integer. Comparison with equation (7.19) shows that we should take n as equal to at least three.

In fact, the transfer function $C(m; -m)$ may be readily calculated for an array of finite sized apertures, for as this is a periodic function its Fourier transform reduces to a Fourier series. The transfer function is also shown in Fig. 7.4 for the case when

$$b = \frac{M\lambda d}{4a} \tag{7.27}$$

for various values of l. For $n = 3$ or 4, imaging approximates reasonably that of a confocal microscope.

This analysis has assumed a one-dimensional geometry, but the results may be extended directly to two dimensions. The transfer functions of Fig. 7.4 apply for imaging of line structures when the aperture discs consist of a square array of square apertures, as then the functions S and D separate into

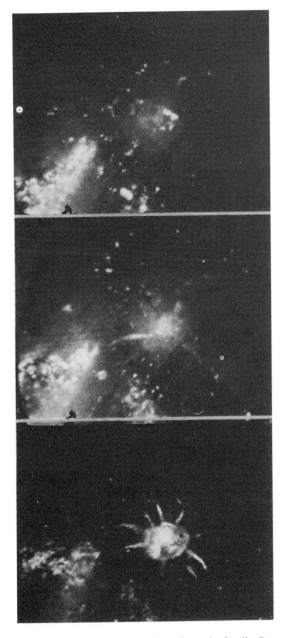

FIG. 7.5. A series of images showing a protozone from the family Coccolithophoridae taken with a direct view scanning microscope incorporating a 60X oil immersion apochromat (numerical aperture 1·0).

products of functions of x and y. The fraction of light used in the image is then as given in the figure, and can be about 10%.

For an array of circular apertures we would expect the effect of aperture size and spacing on the imaging to be substantially similar. We should choose the apertures to be distributed in either a quasi-square or quasi-hexagonal array. The array of holes has to be very regular in order to avoid streakiness in the image. In practice, several hundred apertures would be used around the circumference, the total number of apertures being several tens of thousands. The speed of rotation is governed by the condition that each object element is scanned at least ten or twenty times per second in order to avoid the sensation of flicker.

7.3   A practical direct view scanning microscope

A practical direct view scanning microscope has been successfully constructed by Petran [7.1]. The rotating scanning disc (Nipkow wheel) was 85 mm in diameter and 20 μm thick. The apertures, which were 90 μm in diameter, were arranged in 80 Archimedian spirals to reduce the flicker. The mean nearest distance between openings was 280 μm. In normal operation the disc is rotated 3 times per second, scanning the entire optical field 120 times per second.

We have already seen that the direct view scanning microscope has many of the same imaging properties as the confocal instrument. In particular it is clear that the use of a scanning aperture disc results in rejection of detail which does not come from the focal plane. As a result the direct view microscope has direct application in the examination of thick objects. Figure 7.5 shows a series of images of microfossils in flint taken with a microscope similar to the one of reference [7.1]. The vertical distance between individual pictures is 5 μm and the depth discrimination property is demonstrated.

Reference

[7.1] M. Petráň, M. Hadravsky, M. D. Egger and R. Galambos (1968). *JOSA* **58**, 661.

Chapter 8

The Practical Aspects of Scanning Optical Microscopy

8.1 Introduction

We will now turn our attention to the more practical aspects of scanning optical microscopy and discuss the various criteria which effect the design of a practical instrument. It is clearly not possible to produce a definitive design as this will inevitably depend on the application and types of specimen to be examined.

A scanning optical microscope essentially consists of the following: a light source, objective lenses, photodetector, image processing electronics, some method of scanning the object relative to the focussed light spot. We will now discuss each of these components in turn.

8.2 The light source

In the earliest scanning optical microscopes the scanning spot of a television tube was used as the light source and although this enabled images to be obtained it was far from ideal, the major problem being that the intensity was neither sufficiently stable nor bright. It was probably this lack of light source which was responsible for the relatively slow development of the instrument after the first demonstration of its capabilities in the early fifties.

The renaissance of interest dates from the invention and commercial availability of the laser which provides a very bright light source which can be made sufficiently stable to allow the contrast of images to be enhanced electronically to a very high degree.

The resolution of the microscope depends ultimately on the wavelength of the radiation used and therefore an appropriate laser should be chosen for

169

each application. Unfortunately it is not possible at present to produce a laser with a continuously variable wavelength in all regions of interest: however, discrete lines are available in the range 0·325 to 1·13 μm, and frequency doubling allows shorter wavelengths to be produced. It is also possible to choose a laser of a certain wavelength to excite some process in a specimen such as fluorescence or harmonic generation and then detect the result of this excitation.

We now move on to discuss the optical power which is required to produce an image of adequate quality. It is widely accepted that the signal-to-noise ratio required in the image signal (for observation by the eye) is about five times the ratio of the maximum brightness in the image to the minimum observable variation in brightness [8.1]. As the shot noise due to the quantum nature of light obeys Poisson statistics, we may therefore estimate that for an image with ten grey levels, as is usually adequate for scanning electron microscopy [8.2], we require a signal of about 2500 quanta per picture point. For an image of studio monitor quality we need 36 grey levels, or about 33 000 quanta per picture point. Even two grey levels, or 100 photons per picture point can give an acceptable image. Of course, if we intend to enhance the contrast of the image by n times, we need to increase the number of quanta per picture point by a factor of n^2.

Specimen damage may result either from a quantum effect or from the associated rise in temperature. It is well known that excessive exposure to ultraviolet radiation will result in damage to biological specimens. This is because the radiation is strongly absorbed and causes photopolymerisation. In a scanning microscope, if a cooled photomultiplier is used for detection, the radiation flux may be reduced to a minimum for a given exposure. It should also be remembered that the increased absorption relative to visible wavelengths results in an increase in image contrast.

Heat generated in the specimen diffuses away through itself and also the adjacent microscope or cover slip. The time constant for this diffusion is of the order of 1 μs [8.3]. Let us consider the two limiting cases as the dwell time is much greater or less than the diffusion time constant. For the former case the temperature rise is proportional to the *power* incident on the specimen, and it is estimated that for a 10 K temperature rise an optical beam power of about 35 mW is permissible [8.4].

In the visible region of the spectrum 36 grey levels corresponds to a signal energy of about 10^{-4} J per picture point. For an image produced with line frequently of 100 Hz (which is typical for mechanically scanned systems) we require a beam power of about 10^{-14} W after allowing for a detection efficiency of 10 % and a lower bound on transmission of 10 % through the confocal pinhole. This produces negligible heating of the specimen, and indeed it is seen that we have about 10^5 times as much as we need to produce a high-quality image when beam powers are such that we are in danger of damaging the specimen. With a typical 1-mW laser, assuming that 10% of the light propagates to the specimen (we must expand the beam to illuminate the object approximately uniformly) we still have 10^3 times the power we need to produce a high-quality image, although the power can be reduced using a neutral density filter if necessary. The contrast enhancement that can be obtained is not in practice limited by quantum noise but rather by the intensity fluctuations on the laser beam, unless the image signal is divided by the measured incident laser power.

If the dwell time is much smaller than the diffusion time constant the temperture rise is proportional to the *energy* incident on the picture point as the heat does not have time to escape. Thus at very high scan speeds the temperature rise tends to a limit for constant energy per picture point. It has been estimated that for a maximum temperature rise of 10 K an optical beam energy of about 2×10^{-8} J can be used [8.3], and this energy is about 10^4 times that necessary to produce an image with 36 grey levels.

8.3   Objective lenses

The choice of objectives is to some extent dictated by the scanning technique adopted. However in all cases if a laser light source is employed the illumination is monochromatic and if the lenses are suitably corrected chromatic aberration is not a problem.

If it is chosen to scan the laser beam across a stationary object then very high quality objectives will be needed as the light will be passing through the lens at various angles to the optic axis through a cycle of the scan. Here it will be necessary to choose lenses with very good on-axis and off-axis aberration correction, particularly if confocal operation is required.

The beam scanning method is sometimes used in practice, it being perhaps more common to scan the object relative to a stationary light spot. In theory this relaxes all requirements for the lens to have good (or any!) off-axis aberration correction; the only aberration of importance is spherical aberration which can be compensated to some extent by defocusing. It is quite feasible, therefore, to design and build a very simple lens which will perform quite satisfactorily in such a microscope. This has been demonstrated in the Philips video disc player, which is essentially a scanning optical microscope, where the disc is mechanically scanned relative to a stationary laser beam which is focused into a diffraction limited spot by a very inexpensive pressed plastic aspheric lens [8.5]. It is unlikely that in a practical microscope the optical system will be set up exactly on-axis and so it would be wise to include some correction for the more serious off-axis aberrations such as primary coma.

Unfortunately no such specialised objectives are currently produced commercially and a choice must therefore be made from what is available, a practical experiment being, in our experience, the only reliable way to select the appropriate objective. However it may be feasible with appropriate design to take advantage of the relaxed off-axis aberration requirement to produce an objective which combines extremely long working distance with high numerical aperture, or indeed to improve upon the numerical aperture or minimum operating wavelength wavelength of conventional microscope objectives.

8.4   The photodetector

For the majority of applications a silicon semiconductor photodetector is suitable. If the signal level is very low then a photomultiplier may be used, which in the harmonic microscope may have to be cooled in order to detect the radiation with a reasonable signal-to-noise ratio. Care should be taken in both these cases to ensure that the detector is sensitive to the wavelength it is to detect. A standard silicon photodiode, for example, is ideal for microscopes using He–Ne lasers which have energies substantially greater than the energy gap of silicon but is not suitable for a microscope using long-wavelength ($\gtrsim 1.2\,\mu m$) radiation.

A further design criterion is that the photodiode and its amplifier should have a sufficiently wide bandwidth. This is not a constant but depends on the scanning speed and the number of scan lines in one picture. If we imagine a picture containing n lines with an aspect ratio r then we have rn^2 pixels per frame which corresponds to a bandwidth of $rn^2/2T$ where T is the time taken to scan one picture frame. If we substitute typical figures into this a mechanical scanning system may need a bandwidth of the order of 100 kHz whereas if the scanning were at TV rate the bandwidth would need to be around 10 MHz.

8.5   The image processing electronics

The image processing electronics are substantially similar to those used on scanning electron and scanning acoustic microscopes and in the usual analogue configuration the main processing functions are the gain and black level controls associated with the contrast enhancement facility. An alternative is to digitise the image and store it in a computer allowing much greater flexibility. This suggests the possibility of using various deconvolution techniques to obtain super-resolution, and also of using analysis techniques for particle sizing and sorting. This is discussed in more detail in Chapter 6.

8.6 The scanning system

This is perhaps the most important feature of the design of either scanning optical or scanning acoustic microscopes. There are many methods which are common to both kinds of instruments and therefore we will also include some comments on scanning schemes adopted by acoustic microscopists. In the optical case there are two alternatives: we may either scan the laser beam relative to a stationary object or scan the object relative to a stationary focused spot. These alternatives both have advantages and disadvantages and we will begin by discussing the former arrangement.

8.6.1 *Beam scanning methods*

These have the advantage that the beam may be scanned over the object at high speed and thereby a real-time flicker free picture may be obtained. Images may also be obtained from areas of very large objects or those otherwise unsuitable for mechanical scanning. The beam may be scanned by a variety of means the most popular of which include acousto-optic beam deflectors using chirped gratings, galvonometer mirror-type scanners, rotating multi-faceted mirrors, or mechanical scanning of the objective lens. The advantages and disadvantages of the various methods are summarised in Table 8.1. The figures quoted are for a particular application but are nonetheless useful for comparison.

Beam scanning suffers from some rather serious drawbacks if the miroscope is to operate at its highest resolution. If light is deflected by, for example, a scanning mirror and is incident on an objective then either the full aperture of the lens will not be properly filled at all times during the scan, resulting in the objective having a reduced numerical aperture and a consequent loss of resolution (Fig. 8.1(a)) or alternatively a great proportion of the radiation will be lost (Fig. 8.1(b)). Although both these arrangements suffer from disadvantages they have been used in practice. However by the use of an auxiliary lens (Fig. 8.1(c)) it is possible to ensure that the objective aperture is completely filled without loss of power. The auxiliary lens must of course be of high quality. The rotating mirror must be placed in a zero-deflection plane a distance d_1 in front of the auxiliary lens, where (Fig. 8.1(c))

$$d_1 = (d_2 + d_3)d_2/d_3. \tag{8.1}$$

As there is a specific position in which the scanning mirror must be positioned if mirrors are to be used for deflection in both x and y directions it is preferable to employ a further auxiliary lens to produce a second zero-deflection plane. In practice each auxiliary lens is often replaced by two lenses, the mirror being placed in a parallel beam, in which case d_2 is equal to the focal length of the auxiliary lens.

TABLE 8.1

A comparison of the various beam scanning methods (after V. Wilke, New Technologies in Microscopy, Laser-Scan-Microscope)

Deflector	Maximum scan angle	Number of points	Line-time	Frame-time	Advantages	Disadvantages
Acousto-optical (3 × 10 mm)	2·7	470	15–60 µs	10 ms	Variable, fast, position control	Dispersive, efficiency deflection dependent, complicated optical set-up
Servo control galvanometer (5 mm)	7	610	4 ms	2 s	Variable, good linearity, good position control	Slow, 20% dead time per line
Resonant galvanometer (8 kHz, 5 mm)	7	610	30 µs	64 ms	Variable, fast	Nonlinear scan, long dead-time (90 µs), synchronisation on deflector necessary
Reflecting polygon, 12 facets (1000 rpm)	48	16 700	18 µs	0·25 s	Good linearity	Fixed scan angle, long dead-time (500 µs), hardly variable scan velocity, synchronisation on deflection necessary

In a transmission microscope beam scanning results in vignetting which places a serious limitation on the degree of contrast enhancement that may be used. Unless further scanning mirrors are employed after the light has passed through the object the beam will scan across the detector surface further limiting the available contrast enhancement. A further lens, or lenses,

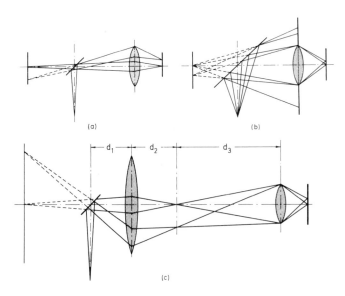

(a)   (b)

d_1 d_2 d_3

(c)

FIG. 8.1. (a) The use of a scanning mirror may result in a reduced numerical aperture of the objective lens. (b) The use of a scanning mirror may result in the loss of a great proportion of the radiation. (c) The use of an auxiliary lens overcomes the problems of (a) and (b).

may be used to minimise this displacement at the detector. It seems unlikely that a transmission confocal microscope could be constructed as it would be necessary to scan the point detector accurately in synchronism with the beam. There is also some doubt as to whether the off-axis correction of a microscope objective is adequate for confocal imaging.

The beam scanning approach has been used quite satisfactorily for relatively low resolution work and in some of the semiconductor applications of Chapter 9. Much of the recently reported high-resolution work, however, has been performed with microscopes using a scanned object and a stationary light beam; it is to the design of these systems that we now turn our attention.

8.6.2   *Object scanning*

A drawback of this approach is that mechanical scanning of the object is inevitably slow (unless the scan area is small) and thereby requires a long persistence display screen. The more recent designs permit an image to be scanned in one or two seconds and thus the image on the screen is built up in a similar time and fashion to the scanning electron microscope.

This approach ensures that cheaper lenses can be used as only the one-axis light has to be dealt with. This also ensures the uniformity of the brightness of the focused spot at all points of the object. A further consequence here is that the image has identical resolution at the extremities of the scan as in the centre. However perhaps the most important advantage is that it is relatively easy to build a confocal microscope by placing a stationary limiting pinhole in front of the photodetector.

We may now make some general comments on the design which apply equally well to acoustic as optical instruments. The specimen movement we wish to obtain must be of very high precision, moving smoothly in the x and y scan directions but being firmly constrained in the focal (z) plane to an accuracy of half a wavelength or better, particularly for confocal or interference microscopy. It is important that the scan be as regular as possible otherwise the picture geometry will be irregular and distorted, while movement in the axial (z) direction will cause a loss in resolution by defocusing.

Although the system must be capable of scanning very small distances accurately for high magnification work the scan amplitude should also be sufficiently high to allow large features to be imaged and to permit the easy location of the area of interest on a large specimen. The design should also possess high enough scan frequencies such that a real time picture with sufficient scan lines may be obtained. Ideally therefore a line scan frequency of several hundred Hertz and a scan amplitude of up to several millimeters would be desirable. However in practice the scanning schemes are usually mechanical and so both of these figures are limited by the maximum acceleration that can be given to the specimen stage. If only high magnification is required the scan speed may be increased substantially, allowing observation of quickly changing specimens.

One very successful mechanical scanning scheme involves clamping the specimen carrying stage between three or four taut piano wires and coupling the corners to electromechanical vibrators or loudspeaker coils to provide the scan. The arrangement has the advantage that the tensions induced in the piano wires during the scan are so directed as to keep the specimen always in the focal plane. Careful design of the coupling linkage between the stage and vibrators together with the correct operating frequencies ensures negligible interference between the line and frame scans. A similar system designed by Brakenhoff [8.6] uses diamond bearings between the stage and a base plate rather than piano wires to locate the focal plane.

These schemes are popular in scanning optical microscopy but in the acoustic case many instruments have independent line and frame scans and often scan the heavy lens in the slow direction and the light object in the fast direction. The independent scanning geometry has several advantages, it permits identical electronics for each channel, allowing easy interchange of fast and slow axes if desired. In particular it lends itself well to digital control. Hollis and Hammer [8.7] have implemented such a scheme with an analogue servo to obtain precise positioning and scanning reproducibility. This enabled them to generate acoustic micrographs made from multiple overlays with excellent registration.

The accurate sensing of position of the stage is an important problem, another solution has been proposed by Heiserman [8.8] who mounts a specimen horizontally on top of a flexible pillar made of small diameter aluminium tubing. Mounted below the sample holder are four small coils of wire spaced at right angles in a plane normal to the pillar with a set of stationary magnets situated about each coil. An orthogonal pair of coils are used to scan the sample while the other two are used to detect the scan velocity. The velocity signals are then used in a servo loop to improve the mechanical response of the system and also integrated to give the position of the sample.

8.7 Object preparation

The techniques employed here are substantially the same as those used in conventional optical microscopy. The only extra point to note is concerned with the transmission confocal microscope. In the confocal microscope both lenses play equal in the imaging and thus we cannot mount a transmission specimen on a microscope slide because unless we are using a very special second collector lens this would introduce aberrations. Instead we must mount transmission specimens between two cover glasses and use cover glass corrected objectives. This precaution is not necessary in the Type 1 scanning microscope where, as we have seen, the aberrations of the collector lens do not affect the imaging.

References

[8.1] A. Rose (1948). *Adv. Electron.* **1**, 131.
[8.2] C. W. Oatley (1972). "The Scanning Optical Microscope", Part 1, The Instrument. Cambridge University Press, Cambridge.

[8.3] C. J. R. Sheppard and R. Kompfner (1978). *Appl. Opt.* **17**, 2879.
[8.4] A. Ashkin and G. D. Boyd (1966). *IEEE J. QE* **QE-2**, 109.
[8.5] G. Bouwhuis and P. Burgstede (1973). *Philips Tech. Rev.* **33**, 186.
[8.6] G. J. Brakenhoff, P. Blom and P. Barends (1979). *J. Microsc.* **117**, 219.
[8.7] R. L. Hollis and R. Hammer (1980). *In* "Scanned Image Microscopy" (Ed. E. A. Ash), p. 155. Academic Press, London and New York.
[8.8] J. Heiserman (1980). *In* "Scanned Image Microscopy" (Ed. E. A. Ash), p. 71. Academic Press, London and New York.

Chapter 9

The Scanning Optical Microscopy of Semiconductors and Semiconducting Devices

9.1 Introduction

In recent years the steadily decreasing size and increasing complexity of semiconductor devices and circuits has necessitated the use of microscopes in their inspection. The scanning electron microscope has been widely used, as, in addition to its high resolution, it may be operated in a variety of contrast modes which give some indication of the electrical performance of the device—information which cannot be obtained by merely examining the specimen surface. The most usual ancillary modes of operation are X-ray analysis, voltage contrast, beam-induced current and cathodoluminescence.

In the X-ray analysis mode we have a high resolution microprobe capable of assessing and measuring material composition. The voltage contrast mode is useful in detecting faults in large integrated circuits. The contrast here arises from variations in the secondary electron signal with the potential difference between various parts of the sample and earth. The beam-induced current and cathodoluminescence modes result from the excess electron-hole pairs generated in the specimen by the incident electron beam. The majority of these carriers form a current which can be measured externally. However, a small fraction, under certain circumstances, recombine radiatively with the resultant emission of radiation (cathodoluminescence). Both of these modes give information about the material properties of the device. An excellent review of these techniques is given by Holt [9.1].

Since the introduction of the laser there has been considerable interest in using laser beams to inject carriers into semiconductor specimens to measure their properties in a fashion analogous to the electron beam of a scanning electron microscope. The initial attractions of the scanning optical micro-cope for this kind of analysis were the elimination of the need for a vacuum system, ease of operation, and low cost. However more recent work has revealed definite advantages to the use of a laser beam such as accurate control of the excitation energy.

9.2   Theoretical background to the OBIC method

We will now discuss the physical principles behind the use of a scanning laser beam to measure the electrical properties of semiconductor devices. The mechanism of carrier generation is indicated in Fig. 9.1 where illumination of the specimen with radiation of energy higher than the band gap causes the generation of excess carriers. Efficient charge collection occurs in materials where the density of excess carriers is greater than or comparable to the equilibrium density in the absence of the laser beam. It is important, therefore, to ascertain whether sufficient carriers can be induced using lasers of sufficiently low power such that when they are focused onto the specimen surface the resulting power density does not damage the specimen in any way.

The rate of photon emission from a laser =

$$\frac{\text{Energy per second}}{\text{Energy per photon}} = \frac{E}{h\nu} \tag{9.1}$$

where E is the laser power, h Planks constant and ν the frequency of the laser radiation.

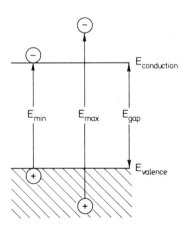

FIG. 9.1. The energy level diagram illustrating the mechanism of carrier generation.

Thus a 1-mW He–Ne (0·6328 Å) laser produces $3 \cdot 2 \times 10^{15}$ photons per second. In practice, not all these photons will reach the specimen: some, for example, will be lost in the optical system of the microscope. We therefore suppose that the beam is attenuated by an amount α such that on striking the semiconductor surface $3 \cdot 2 \times 10^{15}$ α electron–hole pairs are generated per second.

The excess carrier density may now be calculated from

(Number of electron–hole pairs generated)/m^3

$$= \frac{(\text{Number generated per second}) \times \text{lifetime}}{\text{Generation volume}}$$

$$= \Delta n. \tag{9.2}$$

The generation volume is controlled purely by the optical absorption of the beam. In order to obtain a numerical estimate we can crudely approximate this volume as a cone with a base diameter that is determined by the focused laser spot size and the carrier diffusion length, and with a height determined by the beam attenuation and diffusion length (Fig. 9.2).

Typically, we can take spot diameter $= 1 \, \mu m$; diffusion length $= 1 \, \mu m$; absorption length $= 3 \, \mu m$; lifetime $= 2 \times 10^{-3}$ s. This gives

$$\Delta n = 6\alpha \times 10^{23} \, \text{cm}^{-3}. \tag{9.3}$$

Thus if we arrange for a beam attenuation, α, of 10^{-2} we obtain

$$\Delta n \sim 10^{21} \, \text{cm}^{-3}, \tag{9.4}$$

which is comparable to typical equilibrium carrier densities. This estimate is an upper limit for silicon. In practice, the laser beam needs to be much attenuated to provide a suitable number of carriers.

We have now demonstrated that it is possible to induce sufficient excess carriers into a semiconductor specimen with a laser beam. The next problem is how to detect these carriers in an external circuit. In the absence of an external voltage source, the induced electron–hole pairs will only flow

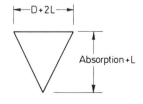

FIG. 9.2. The conical carrier generation volume.

externally if the specimen exhibits photo-voltaic effects (either barrier or bulk) to generate e.m.f's in the specimen and so drive currents around the circuit. This is the case for a *p–n* junction where a built-in voltage due to the space charged depletion region is present. It is possible, therefore, to collect photo-induced currents from a *p–n* junction under no bias.

This effect (Fig. 9.3) can manifest itself as either a maximum current, I_{sc}, but no voltage, if the detector circuit has negligible resistance, or as a maximum voltage, V_{oc}, but no current for very large resistance. For intermediate values of detector resistance intermediate values of current and voltage will be measured. We can therefore see that it is undesirable to speak of this effect as either a photo-voltage or a photo-current.

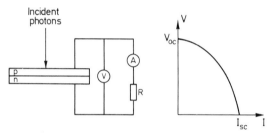

FIG. 9.3. The detection circuit illustrating the role of the external load resistor.

We can understand the semiconductor physics in more detail in terms of the band theory of solids, according to which the process is to raise electrons from the valence band to the conduction band. The energy band diagram of a *p–n* junction is shown in Fig. 9.4. The band bending arises from the requirement that, in equilibrium, the Fermi level must be constant across the junction, and that it is near E_c in *n*-type material but near E_v in *p*-type. This may be understood physically by considering the space charge region of positive and negative charge layers on either side of the junction plane. This region is "depleted" of free charge carriers because the Fermi level there is far from both band edges. This dipole layer produces a change in potential, qV_D, the "diffusion voltage"; it is this change in potential which is manifested as energy band bending and the depletion region constitutes the high resistivity barrier. The analogous distribution in a Schottky barrier is shown in Fig. 9.5.

When the specimen is probed by the laser beam the photo-induced electrons and holes generated near the depletion region are separated by the built-in field. This results in a flow of current generated in and near the junction. The separated holes and electrons cannot be eliminated by current flow-back through the barrier as this contains the non-conducting depletion region whose potential step also opposes this current flow. (On energy band diagrams electrons tend to fall and holes rise as, by convention, the electrostatic potential is plotted increasingly upwards.) These separated carriers produce a non-equilibrium potential difference, V, which, unlike V_D, can lead to a current flow in an external circuit. This is indicated in Fig. 9.6 and is the basis of the optical beam induced contrast (OBIC) method of

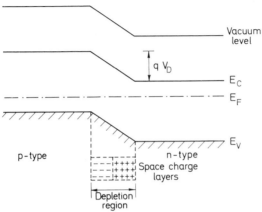

FIG. 9.4. The energy band diagram of a p–n junction illustrating the presence of a depletion region at the junction.

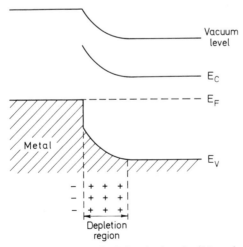

FIG. 9.5. The energy band diagram of a Schottky barrier. Note that the dipole region is a surface charge on the metal and a space charge region extending within the semiconductor. The depletion region in this case lies entirely within the semiconductor.

semiconductor examination. The figure may be interpreted in terms of the Fermi level which is the locus of the free energy of the electrons. As a result of the laser beam excitation the Fermi levels are no longer equal in the p- and n-type material. External work can thus be done, the extra energy coming from the bombarding beam. Figure 9.6 also indicates the effects of varying the resistance of the detection system.

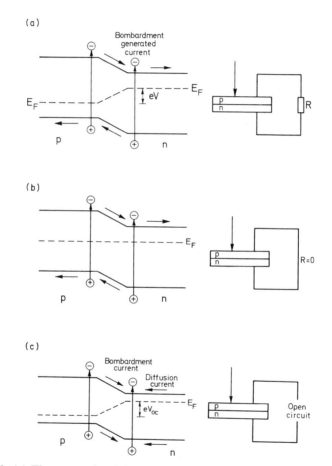

FIG. 9.6. (a) The energy band interpretation of the OBIC process. (b) When the device is short circuited the carriers recombine due to current flow around the external circuit and so no externally detectable voltage is developed. (c) Here there is no external current flow and the electron-hole currents due to beam bombardment and those due to the diffusion of majority carriers over the potential step of the junction are equal and opposite. The potential step is reduced by the open circuit voltage, V_{oc}, which is developed across the junction.

9.3 The OBIC method

The principal problem influencing the efficiency of many semiconductor devices such as a polycrystalline silicon solar cell is the carrier recombination which occurs at planar defects such as grain boundaries near the surface of the material. These defects act as sinks for the photo-excited carriers which would otherwise diffuse to the junction region of the cell and be collected as junction current.

The use of optical scanning has been a popular technique in the investigation of the uniformity of the photo-response over the surface of such semiconductor cells. The early schemes [9.2] were similar to the first scanning optical microscope of Roberts and Young in that a rastered television screen was demagnified and imaged onto the specimen surface. However the low power of the focused spot and the wavelength spread resulted in images with poor resolution and low signal-to-noise ratio. Later workers overcame these problems to a certain extent by using narrow-band light from a monochromator combined with rotating mirror [9.3] or Nipkow disc scanning [9.4] but the resolution was never better than 10 μm and usually a good deal worse.

A laser source, as we have mentioned earlier, is superior to noncoherent sources because it allows a significant photon flux to be focused into a diffraction limited spot. Potter and Sawyer [9.5] scanned a laser beam entering a conventional microscope in order to produce a raster over the device being tested. This, however, has the drawback that for optimum results the scanning mirrors should be placed in the entrance pupil of the microscope which is illuminated with a uniform laser beam [9.6].

These methods, in common with others [9.7–9.9], are similar in that they use ohmic contacts to measure the junction current as the laser beam scans over the sample. Sometimes photovoltages instead of photocurrents are observed. Lile and Davis [9.9] have used a transparent liquid electrode to measure the surface photovoltage. DiStefano and Cuomo [9.10] proposed an alternative scheme involving a capacitance detection technique. This has the advantage that a cell may be examined before the contact wires are attached. This scanned surface photovoltage method relies on a remote electrode capacitatively detecting surface potential fluctuations as the laser beam traverses a defect such as a grain boundary which has a high density of recombination centres. The signal is usually differentiated before being displayed in order to accentuate the signal around a defect. A similar scheme has also been used by Munakata et al. [9.11].

There are considerable advantages to be gained by electing to scan the object rather than the laser beam. The most important is that the optical system is much simpler, which makes it easier to ensure high resolution images. Figures 9.7 and 9.8 show the results of examining an experimental silicon solar cell in a scanning optical microscope [9.12]. The scanning was achieved using the taut piano wire system of the previous chapter and a 5-mW helium–neon laser. The grain boundaries are clearly visible in the photocurrent mode, the contrast across a boundary being typically 5%. If we compare the two images in more detail we see that some of the lines on the reflected light picture do not appear as dark lines on the photocurrent image, e.g. XY. This is because they are not grain boundaries but electrically inert twins.

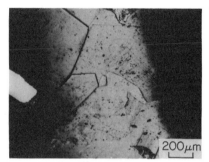

FIG. 9.7. A reflected light image of a pollycrystalline silicon solar cell.

FIG. 9.8. The corresponding OBIC image.

We can now discuss the various dark and bright blobs on the two images. Dark areas which appear dark on both clearly represent dirt of some kind on the surface which is not highly reflective and also prevents light from reaching the cell and producing any photocurrent. The more interesting phenomenon, however, is that of the dark spots in the reflected light image which also appears as bright spots in the photocurrent image. There are two possible explanations for this. First, dirt or drying marks may have contaminated the cell during its fabrication before the final diffusion. These act as masks reducing the surface doping and so reducing the light absorbed in the highly doped surface layer. This is not necessarily a good thing to aim for in a solar cell, as a very highly doped low-carrier lifetime surface layer reduces the lateral surface resistance. Therefore a compromise must be made between high quantum efficiency and low surface resistance in order to obtain maximum power transfer between the cells and the load. The image shown was made with a low load current so that it effectively shows variations in the quantum efficiency of the device. If a large load current were taken, the internal resistance of the cell would affect the contrast, and if the proposed mechanism is indeed the cause of the bright spots they would appear dark with a large load current. An alternative explaination is that there may be some local antireflection action which would be revealed as dark areas in reflection and bright areas in photocurrent.

If the specimen under examination is luminescent, such as a light-emitting diode, we may obtain information complementary to the OBIC method merely by forming an image from the light emitted by the diode. This may be easily achieved merely by scanning a point detector relative to the operating diode and displaying the signal in the usual way [9.13].

9.4 OBIC versus EBIC

The photocurrent mode which we have just discussed is analogous to one which is used in the scanning electron microscope whereby an electron beam is used to inject carriers into the device under inspection. This method has been termed EBIC (electron beam induced contrast), but it is not particularly descriptive of the behaviour occurring. Other terms, such as charge collection, are more appropriate. It is therefore unfortunate but inevitable that the optical case should be termed OBIC. The factors affecting the resolution in each case are clearly different, which makes it important to compare the two techniques.

The comparison may be made by examining the dislocation images from the emitter/base junction of a silicon bipolar transistor which was about 1·8 μm below the specimen surface. For the optical images the carrier generation depth in the specimen corresponds to the absorption length for the incident laser beam, which for helium–neon and silicon is 3 μm. The carrier generation depth in the EBIC case depends on the range of the incident electrons, which for the 30 KeV electrons used to form Fig. 9.10 and silicon is ~ 5 μm. The electron beam diameter was ~ 0·2 μm. Figures 9.9 and 9.10 show the two equivalent high magnification micrographs obtained using beam generated carriers. The OBIC image (Fig. 9.9) reveals dark dislocation lines in a variety of configurations including intersections to give both threefold nodes and small hexagonal arrays. The smallest line widths are ~ 0·5 μm and closely spaced lines can be resolved as separate down to a centre-to-centre line spacing of 1 μm. The dislocations are again revealed as dark lines in the EBIC image of Fig. 9.10 and there is a precise correspondence between these lines and those observed by the OBIC method.

The dark line PQ that can be seen running across the full width of the OBIC image arises because the specimen was previously examined in an SEM to make the EBIC measurements. The electron beam was then scanned along the line PQ for 10 minutes, and this caused a line of contamination to build up on the surface of the SiO_2 film, an effect generally attributed to an interaction between the electron beam, organic vapours present in the vacuum system and the specimen surface.

The OBIC and EBIC images, although obtained by completely different techniques, are very similar to each other. The reasons for this are as follows. Image quality depends on resolution, contrast, and signal-to-noise ratio. For the particular specimen used, the resolution and contrast obtained depends on the lateral spread of the carriers at the depth of the dislocations. This will, in turn, depend on the shape, size and depth of the generation volume, and the subsequent diffusion behaviour of the generated carriers. The generation

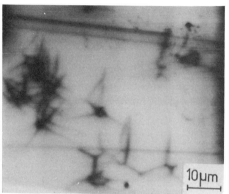

FIG. 9.9. A scanning optical microscope OBIC image revealing dislocations on an experimental silicon bipolar transistor.

FIG. 9.10. The corresponding scanning electron microscope EBIC image.

volume is determined in the SEM by the scattering of the incident beam in the specimen, and in the SOM by the diffraction-limited spot size and the focus position. The limiting factor for the OBIC images is probably the incident beam diameter, and for the EBIC images the scattering of the beam. Nevertheless, for both OBIC and EBIC images and the conditions used, this spread is likely to be $\sim 1\,\mu m$, which agrees with the experimentally observed resolution in both cases of $\sim 1\,\mu m$. For both the OBIC and EBIC images the dislocation contrast was relatively large (1 to 4%) and the collected signal more than adequate. No difficulties due to signal-to-noise problems were encountered [9.14].

In view of the comparable performances of the SOM and SEM for such dislocation studies, it is of interest to consider other possible advantages and disadvantages of using these two types of microscope for such work. The SOM is a much simpler, cheaper instrument which does not require a vacuum system, and the electronic signal processing techniques used in the SEM can be incorporated in the SOM.

The SEM has a smaller minimum probe size, about 0·01 μm, and a very large depth of focus, but in this application the effective probe size is increased owing to the scattering of the electrons inside the specimen. The probe size of the SOM depends on the wavelength used, but could be between 2 and 0.3 μm, and can be maintained below the surface of the specimen.

The depth of penetration in Si [9.15] can be varied continuously over a range of about 0·5 to 60 μm in the SEM by varying the beam energy from 5 to 100 keV. In the SOM the depth depends on the wavelength used, which at present can only be obtained at specific values where a laser line is available. Continuously variable penetration as in the SEM could be obtained in silicon by using a tunable laser. Penetration depths varying from 0·5 to 1000 μm can be obtained in silicon by using available laser wavelengths [9.16] (0·4 to 1·06 μm). The use of a wavelength beyond the band edge affords a means of measuring free carrier absorption, and, therefore, the material doping level [9.17].

The distribution of carriers generated with depth is different for both methods, peaking below the surface for the SEM, and decaying exponentially from the surface for the SOM. Since the carrier lifetime varies with depth owing to surface effects and the varying doping level, some carriers will recombine before reaching the junction and thus not contribute to the conduction current. This will modify the distribution which peaks below the surface for both the SEM and SOM.

There can be charging problems with the SEM due to the surface oxide layers on the device which may have to be removed, and the electron beam in

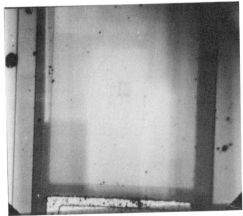

FIG. 9.11. A low magnification reflected light image of the *entire* surface of a silicon transistor which had previously been inspected in a scanning electron microscope. The "tram lines" left by repeated scanning along the same lines are clearly visible.

conjunction with the organic vapours in the vacuum system can deposit lines of contamination on the specimen (Fig. 9.9). This is further illustrated by the *reflection* image of Fig. 9.11. The SOM, however, does not alter or damage the device in this way and can therefore be used as a means of inspecting and testing production devices.

9.5   Scanning extrinsic photocurrent microscopy

The techniques we have considered so far have all been concerned with measuring quantities associated with the minority carrier concentration. Thus, as we have seen, a region of locally high recombination centre density is manifested by a local reduction in the minority carrier lifetime which leads to a reduction in the photocurrent or band-edge luminescence. It has recently become clear that non-radiative recombination centres may be spatially distributed in a very inhomogeneous way. The dark line defect which may be generated during the lifetime of a semiconductor laser or LED is a well-known example. In this case the dislocations grow by a climb process stimulated by non-radiative recombination in the vicinity of the dark-line defect. There are various techniques which can be used to obtain information about such defects [9.18–9.21] but we shall concentrate here on an extension of the junction photocurrent methods by Lang and Henry [9.22].

The microscope system is almost identical to that used for above band gap OBIC, the only difference being that ideally a tunable laser would be incorporated so that the photo-current lineshape could be obtained. The process we are trying to detect occurs in two stages. The first involves excitation from the valence band to the deep level and the second from the deep level to the conduction band. This implies that if the laser photon energy is hv we can only detect deep levels within a region of energy $2hv - E_g$ wide in the centre of the gap (Fig. 9.12). That is, we must have

$$\frac{E_g}{2} < hv < E_g \tag{9.5}$$

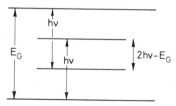

FIG. 9.12. The energy level diagram for the scanning extrinsic photocurrent technique indicating the region of detectable deep level.

where the extremes of $hv = E_g/2$ mean that we could only detect steady-state photocurrent from levels exactly at the middle of the gap, whereas $hv = E_g$ implies that we can detect all deep levels within the gap.

This scanning extrinsic photocurrent (SEP) technique should be seen as complementary to the OBIC method [9.22]. It is possible, for example, for OBIC measurements to indicate reduced current at a microscopic short circuit of the junction or at a location where an opaque defect lies between the junction and the surface thereby blocking the scanning beam. Under these circumstances there should be no SEP signal verifying that the reduced current is not the result of recombination at deep levels. As a further example, when a large concentration of recombination centres reduces the OBIC signal to near zero the SEP technique gives a better measure of the defect concentration as there is essentially no limit to how large the SEP signal may be. We should also finally mention that SEP can form an image of a junction plane located deep within a specimen.

9.6   The measurement of material properties

A traditional technique in the characterisation of semiconductor materials involves the measurement of the decay length of minority carriers generated by a spot of light [9.23]. This is a useful method as the lifetime (τ) in the material may also be found if the diffusion constant is known. It is usual to chop the light spot at a frequency much less than $1/\tau$ to increase the accuracy of the measurement. It is also important that the scanned spot does not come closer than a diffusion length from the crystal boundaries and that the specimen surface is passivated so that the recombination velocity is negligible. The measurements in practice will not be strictly one-dimensional. Thus some corrections are needed particularly near the collecting contact [9.24].

The sensitivity and accuracy of the method may be improved by using an angle lapped junction [9.25]. Hwang *et al.* [9.26], for example, used a 2° angle lap and ultraviolet illumination on a GaP p–n junction and were able to measure diffusion lengths as short as $0.13\,\mu m$ with an accuracy of 15%. They estimated the diameter of the illuminating spot to be of the order of $10\,\mu m$.

One may further take advantage of the angle lapped junction to measure the width of the depletion region as a function of reverse bias and thereby estimate the material doping level. This has been investigated in the scanning electron microscope by MacDonald and Everhart [9.27].

We may also use a chopped beam in our scanning extrinsic photocurrent microscope to measure the electron- and hole-capture or emission cross-sections of the deep energy level impurity or defect centres [9.28]. We can understand the method by considering the physical result of exposing a sample to a chopped light flux with sufficiently long duration. The time constant of the turn on transient depends on the electron and hole emission rates from the defect level, owing to both the multiphonon and photon processes. The turn-off transient on the other hand is governed by multiphonon processes only. Thus the photocurrent transient in the junction obtained by this method gives information on electron and hole emission-capture cross-sections from the defect level for each of these processes.

If we now consider a bulk material we recall that any doping in-homogeneities will produce local gradients in potential causing a charge separation current to flow if excess minority carriers are created in the area. Tauc [9.29] used this photovoltage, produced by a scanning light spot, to measure the resistivity gradient and hence the resistivity of the bulk material. This method has also been used on two-dimensional wafers [9.30] but is slightly unsatisfactory as it is difficult to arrange the sensing electrodes in a configuration such that they do not introduce an apparent inhomogeneity. The bulk photovoltaic measurement also includes detail which is introduced by recombination on crystallographic defects in the material.

Doping inhomogeneities may also be imaged by the use of infrared topography [9.17, 9.31] where a laser with a wavelength which is beyond the band edge is scanned across the sample and the transmitted intensity measured. The variation in doping is then determined from the variation in free carrier absorption. However for good contrast images this technique is best suited to highly doped specimens. The resolution obtained is typically $5\,\mu m$ which should be compared with that attainable by other con-ventional methods. The four-point probe only has a spatial resolution of 1–$2\,mm$, while the best resolution of the spreading resistance method [9.32, 9.33] is $\sim 10\,\mu m$. The disadvantage of these methods is that they damage the semiconductor surfaces by leaving small scratches from the electrical contacts. The infrared topography technique, however, is nondestructive.

A further technique to measure the spatial distribution of doping inhomogeneities is that of electroreflectance [9.34]. Sittig and Zimmermann [9.35] have performed experiments with silicon and shown that an electric field applied perpendicular to the surface of a slice produces a change in reflectance, ΔR, for light in the wavelength range 0.35–$0.38\,\mu m$. The relative change at the fixed wavelength, λ, depends on the magnitude of the electric field as

$$\frac{\Delta R}{R}\bigg|_{\lambda} = r(E). \tag{9.6}$$

If the field is applied to the specimen by constructing a capacitor-like arrangement with the specimen as one of the plates then only variations in the doping concentration, N, should cause a variation in the electric field strength near the surface. That is

$$E = E(N) \tag{9.7}$$

and hence

$$r = r(N) \tag{9.8}$$

where the function r is in general very complicated [9.34]; however for small variations in doping about the mean $\bar{N}$ we linearise $r(N)$ such that

$$\frac{r(x, y) - \bar{r}}{\bar{r}} = \frac{\delta r(x, y)}{\bar{r}} = K \frac{\delta N(x, y)}{\bar{N}} \tag{9.9}$$

where K is a factor depending on $\bar{N}$ and on the band bending at the surface with and without the applied voltage [9.35].

Thus the variation of the electroreflectance signal obtained by scanning a light spot across the surface is directly proportional to the variation of the doping concentration, that is, to the resistivity.

Although electroreflectance spectroscopy is a well-known method of obtaining information on fine details of the band structure it has not replaced photoluminescence measurements. Photoluminescence is a particularly valuable technique in evaluating the performance of many classes of semiconductor devices. It is nondestructive and in the case of a light-emitting diode it gives information on the relative external quantum efficiency and the spectral energy distribution [9.13, 9.36].

It is only relatively recently that the close similarity between electro- and photo-excited luminescence has been confirmed [9.37, 9.38]. Photo-luminescence also gives information on sample purity as the presence of impurities can reduce the efficiency of a luminescent process by providing unwanted radiative (or nonradiative) paths for recombination. The spatial distribution of a luminescent feature will therefore map wafer crystal imperfections and impurity distributions.

Black *et al.* [9.39] were among the first to build a laser scanned photoluminescence microscope. They elected to scan the helium–neon laser beam across the GaAsP sample surface and achieved a spatial resolution of 10 μm. More recently Johnston *et al.* [9.40] have applied the technique to study optically induced degradation in quaternary double heterostructure laser material.

We will now discuss in more detail the design of such a microscope. For the efficient generation of electron-hole pairs the energy of the laser must clearly be greater than that of the energy gap of the semiconductor under study. The signals which have to be detected are usually very small because of the low $(10^{-3}\,\%)$ photoluminescence efficiencies for band to band and impurity recombination processes. Black *et al.* [9.39] used a 5-mW laser and measured a 10^{-6} quantum efficiency from a poor GaAsP sample which means that the detection system must have a rejection ratio of 10^{-7}. This may be relaxed to 10^{-5}, however, if a dark-field arrangement is used. Care must also be taken in determining the scanning speed which must be slow compared to the radiative lifetimes. These are typically 1 ns for band to band transitions and 1 μs for donor–acceptor recombinations [9.41]. We should finally remark that all of the present studies have taken place at room temperature where the photoluminescence data is corrupted owing to phonon scattering. This may be reduced as in cathodoluminescence studies by mounting the specimen on a cooled scanning stage.

9.7   Scanned internal photoemission

We shall now briefly discuss a slightly more specialised technique which is very useful in determining the presence of small inhomogeneities in the contact barrier between two materials, one of which is a semiconductor or an insulator. Photoemission images may also be obtained from Schottky barrier or MOS structures in which the metal is semitransparent. The process of internal photoemission is really a three-step process [9.42]: a photoexcitation of electrons, transport to the surface, and transmission over the barrier. A detailed theory has been given by Williams [9.43].

In practice, scanned internal photoemission images are obtained by scanning a spot of monochromatic [9.44, 9.45] or laser [9.46] light across a metal–insulator–metal sandwich. One of the semitransparent metal electrodes (typically 100–200 Å thick) allows the light to penetrate to the cathode interface. The electrons emitted from the illuminated spot on the cathode are detected as a current which is displayed on a television screen scanned in synchronism with the exciting spot in the usual way.

References

[9.1] D. B. Holt (1974). "Quantitative Scanning Electron Microscopy". Academic Press, New York and London.
[9.2] R. A. Summers (1967). *Solid St. Technol.* **10**, 12 (March 1967).
[9.3] G. Adam (1954). *Physica* **20**, 1037.
[9.4] J. Tihanyi and G. Pazstor (1967). *Solid St. Electron.* **20**, 235.
[9.5] C. N. Potter and D. E. Sawyer (1968). *Rev. Sci. Instrum.* **39**, 180.
[9.6] R. J. Phelan, Jr. and N. L. DeMeo, Jr. (1971). *Appl. Opt.* **10**, 858.
[9.7] J. R. Haberer (1967). *Phys. Failure Eletron.* **5**, 51.
[9.8] J. Tihanyi and G. Pasztor (1967). *Solid St. Electron.* **10**, 235.
[9.9] D. L. Lile and N. M. Davis (1975). *Solid St. Electron.* **18**, 699.
[9.10] T. H. DiStefano and J. J. Cuomo (1977). *Appl. Phys. Lett.* **30**, 351.

[9.11] C. Munakata, M. Nanba and S. Matsubara (1981). *Jap. J. Appl. Phys.* **20**, L137.
[9.12] J. N. Gannaway and T. Wilson (1978). *Electron. Lett.* **14**, 507.
[9.13] T. Wilson, J. N. Gannaway and P. Johnson (1980). *J. Microsc.* **118**, Pt 3, 309.
[9.14] T. Wilson, W. Osicki, J. N. Gannaway and G. R. Booker (1979). *J. Mater. Sci.* **14**, 961.
[9.15] A. F. Makhov (1960). *Sov. Phys. Solid St.* **2**, 1934.
[9.16] W. C. Dash and R. Newman (1955). *Phys. Rev.* **99**, 1151.
[9.17] B. Sherman and J. P. Black (1970). *Appl. Opt.* **9**, 802.
[9.18] B. C. DeLoach, B. W. Hakki, R. L. Hartman and L. A. D'Asaro (1973). *Proc. IEEE*, **61**, 1042.
[9.19] W. D. Johnston and B. I. Miller (1973). *Appl. Phys. Lett.* **23**, 192.
[9.20] W. Heinke and H. J. Queisser (1974). *Phys. Rev. Lett.* **33**, 1082.
[9.21] S. Metz (1977). *Appl. Phys. Lett.* **30**, 296.
[9.22] D. V. Lang and C. H. Henry (1978). *Solid St. Electron.* **21**, 1519.
[9.23] F. S. Goucher (1951). *Phys. Rev.* **81**, 475.
[9.24] I. B. Valdes (1952). *Proc. IRE*, **40**, 1420.
[9.25] K. L. Ashley and J. R. Biard (1967). *IEEE Trans. Electron. Dev.* **ED-14**, 429.
[9.26] C. J. Hwang, S. E. Haszko and A. A. Bergh (1971). *J. Appl. Phys.* **42**, 5117.
[9.27] N. C. MacDonald and T. E. Everhart (1965). *Appl. Phys. Lett.* **7**, 267.
[9.28] C. T. Sah and A. F. Tasch (1967). *Phys. Rev. Lett.* **19**, 69.
[9.29] J. Tauc (1955). *Czech. J. Phys.* **5**, 178.
[9.30] D. L. Blackburn, H. A. Schafft and L. J. Scwarzendruber (1972). *J. Electrochem. Soc.* **119**, 1776.
[9.31] D. C. Gupta and B. Sherman, E. D. Jungbluth and J. F. Black (1971). *Solid St. Technol.* **14**, 44.
[9.32] R. G. Mazur (1967). *J. Electrochem. Soc.* **114**, 255.
[9.33] T. L. Chu and R. L. Ray (1971). *Solid St. Technol.* **14**, 37.
[9.34] M. Cardona (1969). *Solid St. Phys., Suppl.* **11**, 165.
[9.35] R. Sittig and W. Zimmermann (1972). *Phys. Stst. Sol.* **12**, 663.
[9.36] J. F. Black, T. Sentementes and G. Duggan (1972). *J. Electrochem. Soc.* **119**, 369.
[9.37] C. J. Hwang (1967). *J. Appl. Phys.* **38**, 4811.
[9.38] D. R. Scrifes (1971). *Appl. Phys. Lett.* **18**, 160.
[9.39] J. F. Black, C. J. Summers and B. Sherman (1972). *Appl. Opt.* **11**, 1553.
[9.40] W. D. Johnston, G. Y. Epps, R. E. Nahory and M. A. Pollack (1978). *Appl. Phys. Lett.* **33**, 992.
[9.41] J. Vilms and W. E. Spicer (1965). *J. Appl. Phys.* **36**, 2815.
[9.42] C. N. Berglung and W. E. Spicer (1964). *Phys. Rev.* **136**, A1030.
[9.43] R. Williams (1970). *In* "Semiconductors and Semimetals" (Ed. R. K. Willardson), Vol. 6, p. 97. Academic Press, New York and London.
[9.44] T. H. DiStefano and J. M. Viggiano (1974). *IBM J. Res. Dev.* **18**, 94.
[9.45] T. H. DiStefano (1971). *Appl. Phys. Lett.* **19**, 280.
[9.46] R. Williams and M. H. Woods (1972). *J. Appl. Phys.* **43**, 4142.

Chapter 10

Nonlinear Scanning Microscopy

10.1 Introduction

The contrast in conventional microscope images is due primarily to variations in the absorption coefficient and optical thickness of the object under examination. In this case low-level illumination is used. However if the object were probed with a sufficiently intense beam it would behave in a nonlinear fashion and optical harmonics would be produced. The harmonic signal, which is generated in amounts proportional to the nth power of the fundamental intensity, where n is the order of the harmonic, depends on the detailed molecular structure through which the radiation passes. All materials possess third-order nonlinear susceptibilities as well as higher odd order ones. The second-order nonlinear susceptibility is confined to crystals with a structure that exhibit a noncentrosymmetric geometry; such materials are less common but there are sufficient such as $LiNbO_3$, ZnO, $KD*P$ and many biological specimens to be of practical interest. The novelty here is that we are able to study objects using the harmonic radiation generated *within* the object itself.

It is a necessary consequence of this procedure that a form of super-resolution is obtained. If the fundamental radiation has a gaussian distribution it is easy to see that the harmonic radiation will also have a gaussian distribution but with an effective radius $1/\sqrt{n}$ of that of the fundamental. Thus some kind of super-resolution has been obtained, although it might be argued that we could have obtained the same resolution in the conventional way by starting off with the harmonic radiation (however that may have been produced) in the first place. However the important point is that if the image is formed by detecting only the harmonic radiation we produce an image in which the contrast depends on the crystal structure and this image could *not* be obtained in a conventional nonharmonic microscope whatever the

196

frequency of incident radiation used. The wide range of second harmonic generation coefficients [10.1–10.4] should give very strong contrast in the final image.

A further feature of importance is connected with the wide range of nonlinear coefficients of possible object materials. Certain organic molecules, for example, are known to exhibit anomalously large third-order nonlinear polarisabilities. We may also observe the presence of permanent polarisation, induced polarisation and strains in materials by way of the second-order nonlinear susceptibilities they produce.

Since the first observation of second-order harmonic generation (SHG) by Franken et al. [10.5] in 1961 there have been several studies which indicate the potential of a second harmonic microscope. An analysis of the intensity and angular distribution of the second harmonic generated within ferroelectric crystals enabled Miller [10.6], Freund [10.7] and Freund and Hopf [10.8] to determine the properties of ferroelectric domains. Arguello et al. [10.9] have also measured second harmonic generation to find the layer thickness in polytype ZnS. Hellwarth and Christensen [10.10] developed one of the first second harmonic microscopes to study more irregular microstructures and were able to observe individual grain shapes, orientations and growth directions in polycrystalline ZnSe "window" material, these features being completely invisible with conventional microscopy.

The Hellworth–Christensen microscope might be called a conventional harmonic microscope in that the *whole* specimen is illuminated with a laser beam and an image formed by focusing the emitted harmonic light. It is also possible to conceive of a scanning harmonic microscope in which a highly focused laser spot is scanned over the specimen and the harmonic signal detected point by point. An acoustic nonlinear microscope of this later form has also been constructed [10.11].

10.2 Imaging in the harmonic microscope

The majority of harmonic microscopes built to date have been concerned with detecting the second harmonic radiation generated within the object and so in the following we shall restrict ourselves to noncentrosymmetric molecules for which the second harmonic generation coefficient d_{ijk} is defined by the equation

$$p_i^{2\omega} = d_{ijk}E_j^{\omega}E_k^{\omega} \tag{10.1}$$

where E_j and E_k are electric field strengths at radian frequency ω in the two directions j and k, and $p_i^{2\omega}$ is the resulting nonlinear polarisation at radian frequency 2ω and in direction i. The second harmonic generation coefficients are very small (of the order of 10^{-23} MKS units), but they do vary over several orders of magnitude for different materials, thus enabling good contrast to be achived.

We may use a very simple model based on plane wave propagation to write an expression for the second harmonic power $P_{2\omega}$ generated in a crystal as [10.12]

$$P_{2\omega} = \frac{2\omega^2}{A} \left(\frac{Z_0}{\sqrt{\varepsilon_r}}\right)^3 \left(\frac{\sin \frac{1}{2}\Delta kl}{\frac{1}{2}\Delta kl}\right)^2 l^2 d^2 P_\omega^2 \tag{10.2}$$

where A is the area of the focused spot, $Z_0 = 120\pi$ is the impedance of free space, ε_r is the relative permittivity of the medium, l is the interaction length, P_ω is the fundamental input power and $\Delta k = k_{2\omega} - 2k_\omega$ is the wave number difference which takes into account the different speeds at which the fundamental and harmonic travel within the specimen. Here we have dropped the suffices on the generation coefficient d.

It is possible to obtain very high harmonic generation coefficients, of the order of 50% with frequency doubling crystals used in pulsed lasers. These are achieved by phase-matching which increases the length over which the fundamental beam interacts usefully with the crystal. Provided that there is a direction in the crystal for which the harmonic and fundamental travel at the same speed, the harmonic generated adds up in phase, thus greatly increasing the efficiency. This corresponds to $\Delta k = 0$. This technique of choosing the specific direction to ensure phase matching clearly cannot be used in a microscope as one wishes to examine small areas of the specimens, and hence very high power densities of the order of 10^{11} W m^{-2} are necessary to produce a reasonable harmonic power. The scanning approach has the great advantage that a much lower laser power is needed to achieve a given power density in the specimen than for the conventional harmonic instrument. Furthermore the fast scanning also results in a lower temperature rise in the specimen for a given harmonic signal; this has immediate advantages when producing images from biological specimens which are very delicate and hence beam heating must be kept to a very low level.

The contrast in the images is due to both the spatial variations in the harmonic generation within the object and also by variations the absorption of the fundamental and harmonic radiation. In the following we will confine our attention to the former of these two mechanisms and examine the imaging properties of conventional and scanning harmonic microscopes in more detail.

To simplify the analysis we restrict ourselves to specimens whose length is small compared with the aperture length [10.13], that is we neglect double refraction and the extraordinary wave nature of the harmonic beam. We may therefore write the second harmonic light emerging from the specimen as

$$E_{\text{harmonic}} = t_2 E^2 \tag{10.3}$$

where E is the fundamental field and t_2 the second harmonic generation coefficient.

If we set such a specimen in the idealised scanning microscope of Fig. 10.1 and arrange to collect the second harmonic light we can use the methods of Chapters 2 and 3 to write the image intensity as [10.14]

$$I(x_s) = \int\limits_{-\infty}^{+\infty}\int\int\int S^2(x_1)h_1^2\left(\frac{x_3 + x_1/M}{\lambda_1 f}\right)h_1^{*2}\left(\frac{x_3' + x_1/M}{\lambda_1 f}\right)$$

$$\times\, t_2(x_3 - x_s)t_2^*(x_3 - x_s')h_2\left(\frac{x_3 + x_s/M}{\lambda_2 f}\right)h_2^*\left(\frac{x_3' + x_s/M}{\lambda_2 f}\right)$$

$$\times\, D(x_s)\,dx_1\,dx_3\,dx_3'\,dx_s \qquad (10.4)$$

where h_1, h_2 are the impulse responses of the lenses, i.e. the Fourier transforms of the pupil functions, M is the spot demagnification d/f, λ_1 is the wavelength of the fundamental light and λ_2 that of the second harmonic.

If we consider the image of a single point object

$$t(x_3 - x_s) = \delta(x_3 - x_s) \qquad (10.5)$$

then

$$I(x_s) = \left\{S^2(Mx_s) \otimes \left|h_1^2\left(\frac{x_s}{\lambda_1 f}\right)\right|^2\right\}$$

$$\times \left\{D(Mx_s) \otimes \left|h_2\left(\frac{x_s}{\lambda_2 f}\right)\right|^2\right\} \qquad (10.6)$$

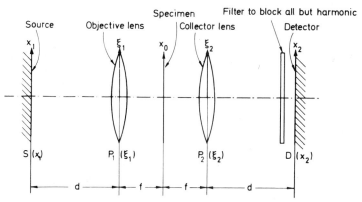

FIG. 10.1. The notation and optical configuration of a scanning harmonic microscope.

where $\otimes$ denotes the convolution operation. The image here is clearly sharpest when both S and D are points and is degraded when either or both have finite size.

An estimate of the resolution attainable can be made by assuming the pupil functions to be equal gaussians. Then the impulse responses may be written

$$h_1 = \exp -\frac{x^2}{a^2}; \quad h_2 = \exp -\frac{4x^2}{a^2}. \tag{10.7}$$

We follow the nomenclature of conventional nonharmonic microscopy of Chapter 3 and denote a scanning microscope with a point source and large, incoherent detector a Type 1 microscope whereas an instrument having both a point source and detector will again be termed a confocal microscope. The behaviour of a conventional harmonic microscope may also be obtained from this generalised model with an incoherent source and point detector.

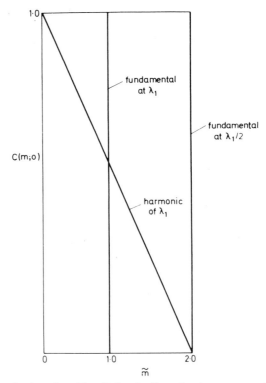

FIG. 10.2. The transfer function $C(m; 0)$ for the Type 1 coherent scanning harmonic microscope.

Thus the intensities in the harmonic images may be written

$$I_{\text{Type 1, harmonic}} = \exp - \frac{4x^2}{a^2},$$

$$I_{\text{conventional, harmonic}} = \exp - \frac{8x^2}{a^2},$$

$$I_{\text{confocal, harmonic}} = \exp - \frac{12x^2}{a^2}. \tag{10.8}$$

Returning to the case of general pupil functions we can now consider the imaging in terms of the Fourier transforms of $t_2(x)$ and as in the previous chapters compare the imaging performance of different geometries by examining the form of the function $C(m; p)$.

If we consider an idealised object

$$t_2(x) = 1 + \alpha \cos 2\pi v x; \quad \alpha \text{ small} \tag{10.9}$$

then

$$I(x_s) = 1 + 2\alpha C(v; 0) \cos 2\pi v x_s. \tag{10.10}$$

that is the imaging is characterised by the function $C(v; 0)$. This function is plotted in Figs 10.2 to 10.4 for the various cases of the Type 1, coherent

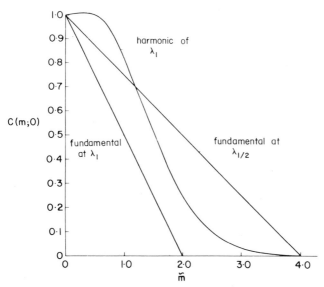

FIG. 10.3. The transfer function $C(m; 0)$ for the Type 1 scanning harmonic microscope.

microscope (small P_2), the Type 1 microscope with collector lens larger or equal to the objective lens and confocal with equal lenses. In all cases it is seen that the high spatial frequency response is inferior to that of a microscope of a similar type illuminated with radiation of the harmonic frequency.

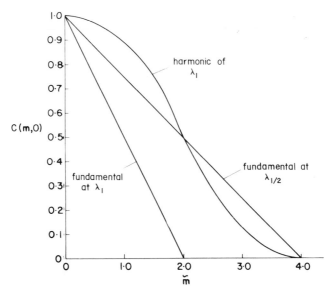

FIG. 10.4. The transfer function $C(m;0)$ for the confocal scanning harmonic microscope.

The result for a conventional harmonic microscope with incoherent source and equal aperture condenser and objective lenses is shown in Fig. 10.5 and it is seen that the high spatial frequency response is better than for a Type 1 microscope, but not as good as for the confocal microscope. We have just discussed the properties of a conventional harmonic microscope with incoherent source but in practice [10.10] such microscopes use laser illumination and hence the source is coherent. If a large area of the specimen is illuminated the system is equivalent to one with a small condenser aperture and then the spatial frequency response becomes identical to a coherent microscope operating at the harmonic frequency, that is the cut-off frequency is half that of a scanning harmonic microscope with equal lens pupils.

The confocal scanning microscope gives superior resolution but this is obtained at the expense of throwing away a large part of the signal. As signal levels are extremely small in the harmonic microscope at present the improvement in resolution does not justify making them any smaller. Thus the preferred approach is the use of a Type 1 scanning microscope with equal lens pupils.

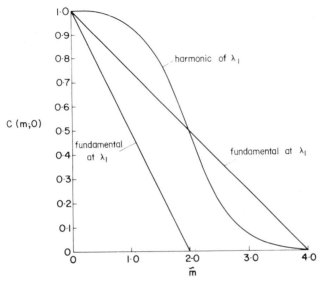

FIG. 10.5. The transfer function $C(m; 0)$ for the conventional harmonic microscope.

10.3   Practical harmonic microscopy

An experimental arrangement of a Type 1 scanning harmonic microscope [10.15–10.16] is shown in Fig. 10.6. The required fundamental power density is achieved in this particular case by focusing a 0·5 W CW Nd-YAG laser beam (wavelength 1·06 μm) into an almost diffraction limited spot which with suitable specimens results in the generation of between 10^{-13} and 10^{-9} W of harmonic power. A cooled photomultiplier tube is needed to detect the signal.

As the harmonic power is very low and must be measured against the large background of the fundamental it is crucial to take great care in filtering to ensure that only the harmonic signal is detected by the photocathode. Firstly it is essential that the specimen is illuminated with only 1.06 μm radiation. This may be achieved by using a GaAs band edge filter. With correct choice of the doping level the band edge can form a very high performance low pass filter with a low insertion loss at 1·06 μm and a loss of hundreds of decibels at all higher frequencies. This scheme effectively removes all the visible pump light and harmonics of 1·06 μm from the laser beam. The narrow-band interference filters in front of the photomultiplier tube, which each have a transmission of 50% at the harmonic frequency and a loss of at least 40 dB at all other frequencies, serve to remove most of the fundamental, stray room light, and also any fluorescence or thermal radiation from the specimen. The rapidly falling response of the S20 photocathode in the infrared further

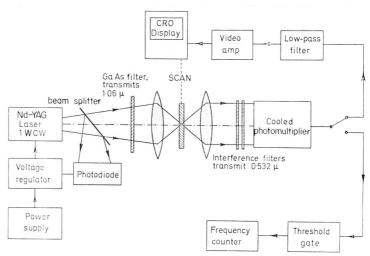

FIG. 10.6. A schematic arrangement of a Type 1 scanning harmonic microscope.

attenuates the fundamental signal. Since the power density is only appreciable in the focal region there will be no detectable harmonics generated by any of the filters or objectives in the system.

An alternative approach to the problem of detecting the very low harmonic signal levels is the use of heterodyne detection. A portion of the incident beam would bypass the object and pass firstly through a frequency doubling crystal and then a frequency shifting element such that if the frequency of the fundamental beam were ω the reference beam would then be at a frequency of $2\omega + \Delta\omega$. This would then mix with the signal from the object consisting of the fundamental and required harmonic radiation. The final field would be of the form

$$U = A \exp j\omega t + B \exp j2\omega t + C \exp j(2\omega + \Delta\omega)t \qquad (10.13)$$

where A is the fundamental signal, B is the harmonic and C the reference. Following Yariv [10.17] we can write the signal detected by a photomultiplier as $|U|^2$ and if we also arrange to collect only the signal at a frequency $\Delta\omega$ we have

$$I_{\text{detected}} \sim 2 \, \text{Re} \, \{C^*B\} \cos \Delta\omega t \qquad (10.14)$$

and so by this method we are able to increase the detected signal level of the harmonic beam.

The choice of laser for such experiments is somewhat arbitrary but would be governed by the transmission characteristics of the available filters and specimens. It is also possible to decrease the temperature rise in the specimen by increasing the scanning speed. This results in the whole specimen reaching an equilibrium temperature, in contrast to the very slow scan case where the probe is essentially stationary on the object, and also allows a high incident power to be used. The heating effect may also be minimised by choosing a specimen with low absorption at the fundamental, although it is perhaps equally desirable to have low loss at the harmonic frequency to avoid attenuation of the harmonic signal escaping from the crystal. This harmonic attentuation can be avoided by focusing the beam on the far side of the specimen but the price one pays for this is the increase in spot size due to the spherical aberration introduced.

A further important practical point derives from harmonic generation being a square law process such that any variations in the laser output power produce larger changes in the harmonic signal. It may therefore be necessary to run the pump lamp of the laser from a stabilised d.c. supply and to use feedback from the fundamental beam to control the lamp voltage (Fig. 10.6).

The majority of specimens used in feasibility studies in harmonic microscopy have been very transparent, high melting point crystals which were not damaged by the high power densities to which they were subjected. Figure 10.7, for example, shows two second harmonic images from a roughly

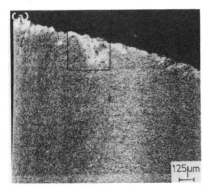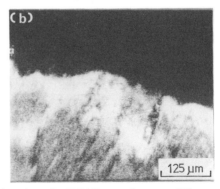

FIG. 10.7. Second harmonic pictures of the edge of a KD*P crystal at two different magnifications. The rectangle in (a) shows the field of view of (b).

polished KD*P specimen. The area in the rectangle in (a) shows the field of view in (b). The signal-to-noise ratio is barely adequate here as quantum noise can be seen; however it is clear that the contrast is greater at the edge of the crystal. This is to be expected as the lattice is likely to have been distorted in this region when the sample was cut, thereby leading to widely varying values of the second harmonic generation coefficient. Figure 10.8 shows comparison conventional and second harmonic micrographs of a sample of lithium niobate; the scanning geometry is not perfect in this case but there is considerable detail present in the harmonic picture, such as the diagonal shading which is not present in the conventional image.

 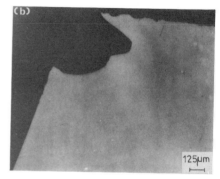

FIG. 10.8. The polished surface of a lithium niobate crystal: (a) second harmonic picture; (b) as seen under a conventional microscope.

The depth of focus is of great importance in harmonic microscopy. Detail outside the focal plane does not interfere with the image as much as in conventional microscopy since the harmonic generated is proportional to the intensity squared and this results in the main contribution only coming from the region of focus where the intensity is very large. This effect is illustrated in Fig. 10.9. Pictures can be selected from any depth of the crystal although as mentioned previously spherical aberration may reduce the resolution. This lithium niobate specimen consisted of a thin slab, visible on the far right of the micrographs, with an inclined face on the left-hand part of the pictures.

10.4   The future of nonlinear microscopy

Although second harmonic microscopy of the kind we have just described is still in its infancy there is another technique which would allow the scanning nonlinear optical microscope to attain contrast which is enormously greater

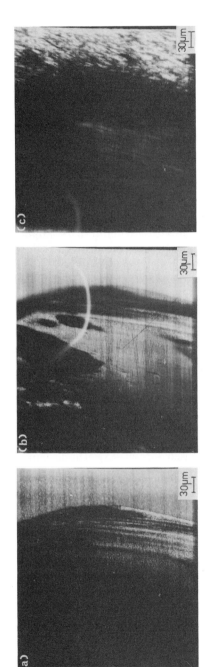

FIG. 10.9. Second harmonic pictures of a lithium niobate crystal at high magnifications. The laser is focused on (a) the front, (b) the middle and (c) the back of the crystal.

than that of any conventional instrument. The technique is also powerful in that it depends on the third-order nonlinear susceptibility $\chi^{(3)}$ and so applies, in principle, to all crystals and not just those lacking inversion symmetry as in the second-order case. The method makes use of Raman-type resonances and is in some sense similar to coherent anti-Stokes Raman spectroscopy (CARS). The principle is that two optical beams of frequencies ω_1 and ω_2 will be incident on the sample. Via the third-order nonlinear polarisability a collimated, colinear beam will be regenerated at the frequency $2\omega_1 - \omega_2$. Typically we would arrange to fix ω_1 and vary ω_2. When $2\omega_1 - \omega_2$ equals the frequency of a Raman mode (ω_r), $\chi^{(3)}$ is resonantly enhanced and a greatly increased output signal is generated.

The potential importance of this technique arises since it is likely that different portions of a biological specimen, for instance the cell walls versus the cell nucleus, will contain different concentrations of molecules having particular chemical bonds. When the difference of the two incident frequencies, $\omega_1 - \omega_2$, equals the vibrational frequency of the selected bond, only cells containing that bond will radiate. The optical picture obtained will thus be of molecules containing the selected bond. It is immediately clear that this technique will offer orders of magnitude increase in contrast as compared to, for example, a technique based on optical density or phase gradient. It is also of particular importance that though the characteristic bonds are in the infrared no infrared source is required. For instance, if ω_1 is green and ω_2 is red, the generated frequency $2\omega_1 - \omega_2$ will typically be in the blue–green.

A general scheme for this technique consists of a frequency doubled Nd-YAG, argon or krypton ion laser to give the fixed frequency ω_1. This same frequency is then used to pump a dye laser to provide the tuneable ω_2. These two beams are focused onto the object and an image formed by detecting $2\omega_1 - \omega_2$ as described above. Using two dye lasers to give two tuneable beams would result in further controllability.

References

[10.1] D. J. Bradley, M. R. H. Hutchinson and H. Koetser (1972). *Proc. R. Soc.* **A329**, 105.
[10.2] K. C. Rustagi and J. Ducuing (1974). *Opt. Commun.* **10**, 258.
[10.3] J. P. Herman and J. Ducuing (1974). *J. Appl. Phys.* **45**, 5100.
[10.4] B. F. Levine and C. G. Bethea (1975). *J. Chem. Phys.* **63**, 2666.
[10.5] P. A. Franken, A. E. Hill, C. W. Peters and G. Weinreich (1961). *Phys. Rev. Lett.* **7**, 118.
[10.6] R. C. Miller (1964). *Phys. Rev.* **134**, A1313.
[10.7] I. Freund (1967). *Phys. Rev. Lett.* **19**, 1288.
[10.8] I. Freund and L. Hopf (1970). *Phys. Rev. Lett.* **24**, 1017.

[10.9] C. A. Arguello, J. Huzart, G. Mendes, F. Bellon and R. C. C. Leite (1972). *In* "Proceedings of 11th International Conference on the Physics of Semiconductors", pp. 1358–1362. PWN-Polish Scientific Publishers, Warsaw.

[10.10] R. Hellwarth and P. Christensen (1974). *Opt. Commun.* **12**, 318.

[10.11] R. Kompfner and R. A. Lemons (1976). *Appl. Phys. Lett.* **28**, 295.

[10.12] A. Yariv (1967). "Quantum Electronics", 1st edition. Wiley, New York.

[10.13] D. A. Kleinman (1962). *Phys. Rev.* **128**, 1761.

[10.14] T. Wilson and C. J. R. Sheppard (1979). *Opt. Acta* **26**, 761.

[10.15] J. N. Gannaway and C. J. R. Sheppard (1978). *Opt. Quantum Electron.* **10**, 435.

[10.16] J. N. Gannaway and T. Wilson (1979). *Proc. R. Microsc. Soc.* **14**, 170.

[10.17] A. Yariv (1971) "Introduction to Quantum Electronics", 2nd edition. Holt, Rinehart and Winston, New York.

Index